IMAGINARY
WEAPONS

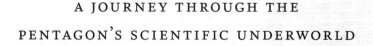

A JOURNEY THROUGH THE
PENTAGON'S SCIENTIFIC UNDERWORLD

SHARON WEINBERGER

NATION BOOKS
NEW YORK

IMAGINARY WEAPONS:
A *Journey Through the Pentagon's Scientific Underworld*

Copyright © 2006 Sharon Weinberger

Published by
Nation Books
An Imprint of Avalon Publishing Group, Inc.
245 West 17th Street, 11th Floor
New York, NY 10011

AVALON
publishing group incorporated

Nation Books is a co-publishing venture of the
Nation Institute and Avalon Publishing Group Incorporated.

Library of Congress Cataloging-in-Publication Data is available.

ISBN-10: 1-56025-849-7
ISBN-13: 978-1-56025-849-0

9 8 7 6 5 4 3 2 1

Book design by India Amos, Neuwirth and Associates, Inc.
Printed in the United States of America
Distributed by Publishers Group West

For Nathan

CONTENTS

USEFUL TERMS AND CAST OF CHARACTERS

USEFUL TERMS

CHAIN REACTION: In a nuclear bomb, the neutrons released during fission cause other nuclei to split apart, sparking a runaway chain reaction. Similarly, a theoretical isomer bomb would need not only a way to "trigger" isomers, but also a mechanism to spark a similar self-sustaining reaction that quickly triggers other isomers. See isomer triggering.

COLD FUSION: The generic term dating back to 1989 associated with the highly controversial theory that nuclear reactions can be generated at room temperature using relatively simple equipment.

DARPA: The Defense Advanced Research Projects Agency, a division of the Pentagon responsible for funding far-out science and technology.

HAFNIUM (HF): A chemical element with the atomic number 72 on the periodic table. A naturally occurring silvery-gray metal, hafnium is used in applications ranging from incandescent lamps to nuclear control rods.

HAFNIUM BOMB: The notional idea of building a bomb based on hafnium-178m2. See also isomer bomb.

HAFNIUM ISOMER: A generic term that covers "isomeric" forms of the element hafnium. In the case of this book, it refers specifically to hafnium-178m2, or the second meta-stable state of hafnium-178. See also isomer.

ISOMER: In this book, isomer refers specifically to a nuclear isomer, an excited, or "charged-up" state of a nucleus. However, there are also chemical and atomic isomers.

ISOTOPE: A form of an element that has the same atomic number, but with a different atomic mass. For example, uranium-238 and uranium-235 are isotopes of the element uranium.

ISOMER BOMB: The notional concept of a weapon with an extremely large explosive force based on nuclear isomers. An isomer weapon could be a bomb or a laser.

ISOMER TRIGGERING (OR HAFNIUM TRIGGERING): The notional concept that a relatively small spark, such as low-energy photons from a dental X-ray machine, could cause nuclear isomers to "de-excite," or release their energy.

GAMMA-RAY LASER (OR GRASER): A notional concept of a laser that uses nuclear isomers as the lasing medium.

HALF-LIFE: The time required for a nucleus to lose half of its radioactive energy. The isomer hafnium-178m2, for example, has a half-life of thirty-one years.

JASON: Typically spelled in capital letters, JASON is a secretive group of elite scientists who advise the government on national security. Also referred to collectively as the JASONs.

NUCLEAR FISSION: The process by which a nucleus splits in two, releasing energy and radioactive byproducts. Fission is the basis of the atomic bomb. In a thermonuclear weapon, or hydrogen bomb, nuclear fission is the trigger, or "primary," that sparks a secondary fusion reaction.

NUCLEAR FUSION: The process of joining two nuclei, which is accompanied by a large release of energy. In a thermonuclear weapon, fusion is the secondary stage of the explosion.

STAR WARS: Popular term—typical derogatory—given to

the ballistic missile defense shield suggested by President Ronald Reagan in the 1980s. The formal name was the Strategic Defense Initiative.

HAFNIUM BELIEVERS

FRED AMBROSE: Intelligence analyst associated with the Defense Intelligence Agency and Central Intelligence Agency, and long-time believer in the isomer bomb.

CARL COLLINS: Professor at the University of Texas at Dallas and lead scientist in the 1998 dental X-ray experiment that prompted many to believe in the isomer bomb.

JAMES COREY: Sandia intelligence analyst with a penchant for fringe science. An early alarmist about foreign isomer bombs. Currently retired in Missouri and running a cold fusion company.

HILL ROBERTS: A private supplier of hafnium isomer and collaborator on the dental X-ray experiment. Also founder of "Lord, I Believe."

EHSAN KHAN: Department of Energy science official and occasional government point man on fringe science. Temporarily serving at DARPA to help manage research into the isomer bomb.

PATRICK McDANIEL: Physicist at Sandia National Laboratories and former Air Force scientist. Principal ally of Carl Collins and a collaborator on the 1998 dental X-ray experiment.

MARTIN STICKLEY: A Pentagon official in charge of the isomer bomb and primary hafnium believer. Currently rumored to be supporting cold fusion research.

TOM WARD: Former official in the National Nuclear Security Administration. Now in private business, Ward works as an organizer of the Hafnium Isomer Production Panel

TONY TETHER: The director of DARPA and proponent of the isomer bomb.

HAFNIUM ENEMIES

JOHN BECKER: Experimental physicist at Lawrence Livermore National Laboratory and lead scientist on two "hafnium triggering" experiments conducted at Argonne National Laboratory in Illinois that attempted, unsuccessfully, to replicate Carl Collins's dental X-ray results.

DONALD GEMMELL: Physicist at the Argonne National Laboratory and a scientist involved in the Argonne counter-experiment.

WILLIAM HERRMANNSFELDT: Physicist at the Stanford Linear Accelerator Center, a Department of Energy National Laboratory. Recruited by Tom Ward to serve on the Hafnium Isomer Production Panel.

STEVEN KOONIN: Physicist at California Institute of Technology and member of the JASONs. Participated in the JASONs' 1999 review of the "hafnium bomb" concept.

JOHN SCHIFFER: Experimental physicist at Argonne National Laboratory and participant in the Argonne counter-experiment

MORT WEISS: Theoretical physicist at the Lawrence Livermore National Laboratory. Once worked on the concept of an isomer weapon.

JERRY WILHELMY: Experimental physicist at the Los Alamos National Laboratory and participant in the Argonne counter-experiment.

PETER ZIMMERMAN: Former chief scientist at the Arms Control and Disarmament Agency. Commissioned JASON study on the hafnium bomb.

RELATED AND OCCASIONAL CHARACTERS

JAMES CARROLL: Physics professor at Youngstown State University and former student of Carl Collins. A coauthor on the original "dental X-ray" paper, Carroll eventually becomes a critic of hafnium triggering.

IOAN-IOVITZ AND ELENA POPESCU: Original Collins's collaborators from Romania. Elena, who was passed over for the Nobel Prize four times (by her husband's account at least), died in 2002.

LEV RIVLIN: A Russian scientist and editor of *Journal of Laser Physics*. Rivlin obtained the first patent for a gamma-ray laser.

PAUL ROBINSON: Co-chair of the Hafnium Isomer Production Panel and one-time director of Sandia National Laboratories.

GEORGE ULLRICH: Former senior Defense Department official in charge of weapons research; investigated the possibility of isomer weapons.

There's a big gap between old-fashioned black powder and a nuclear weapon.

—JAMES TEGNELIA,
Sandia National Laboratories,
March 2003

The Gateway

STANDING IN THE staff kitchen of *Aviation Week & Space Technology* magazine, Amy Butler and I stared at the design for the world's first—and, so far as we knew, only—acoustic weapon.

It was drawn on a paper towel.

The weapon's creator, Franz Gayl, a Marine Corps science advisor, stood next to it and smiled proudly. Franz was on leave from his official Pentagon duties to attend classes in Washington, D.C., at the National Defense University. He had stopped off at our downtown office that summer morning in 2005 to chat about space travel with Amy, the magazine's military editor. Franz was promoting a somewhat offbeat concept that was making the rounds in the Pentagon: a "space plane" that would loft a squad of Marines into orbit—and return them through the atmosphere, ready to kick down doors anywhere on the planet.

The three of us had walked back to the kitchen to take a coffee break, and somewhere along the way, our conversation shifted to an even more unusual topic, Franz's homemade "directed energy" weapon.

Aviation Week writers enjoy swapping stories about men who write to the magazine with strange ideas, ranging from notions of tapping unlimited energy from a quantum vacuum to schematics for newfangled space ships. Patients from mental hospitals would send the editors sketches of elaborate aircraft designs; inmates at state correctional institutions displayed a similar penchant for aeronautics. Sometimes, these would-be Edisons even turn up at the office, ready to demonstrate detailed plans for their antigravity devices. But the major difference here was that Franz wasn't living in a trailer court near Roswell, New Mexico. He actually worked for the Pentagon, and his idea, while certainly farfetched, was not some lunatic's pipe dream.

I had met Franz just six months prior at a military conference in northern Virginia, where he was presenting a slide show on the future of weapons that use light and sound. He effused a sort of intense and heartfelt enthusiasm, and his lecture fit in well at the forum, where defense industry representatives and Pentagon officials were hawking directed-energy weapons: guns that could shoot lightning, devices that transmit the voice of God into your head, among other novelties. Warfare tends to encourage creative thinking, but compared to space planes, Franz's acoustic weapon looked a bit odd—especially as sketched out on a paper towel. He had actually built the weapon, or at least so he told us. The question was: did it work?

Acoustic weapons are based on the notion that a focused blast of sound waves can somehow disorient and deter an aggressor. This is fundamentally different than something that is simply really loud and annoying, like a bullhorn. Scientists have theorized that anything from nausea to immobilizing pain might be a potential effect of sound waves tuned to precise frequencies. For many years, physicists have studied the theoretical basis for such a device, but have largely concluded that such weapons are not really practical. But Franz who, despite his title, was not really a scientist, was possessed by the idea.

He built his weapon, appropriately dubbed the Gayl Blaster, with the help of a New Hampshire company called

Information Unlimited. It was a company I had heard of, though I had to rack my brain to remember where.

"How do you know them?" I asked, squinting at the paper towel schematic.

"When I was at the Naval Postgraduate School, I had them build the phased acoustic device," he said. "I can't do the electronic piece. I understood the theory."

"You called them up?" I asked.

"Yeah, they built it, and the government gave them money," Franz said. "The Naval Postgraduate School gave me research money from the Office of Naval Research and I paid them for it."

Then I remembered. I had once stumbled across Information Unlimited's Web site, which at first glance resembled an online adult novelty shop. In addition to a line of plasma balls and laser pointers, Information Unlimited advertised an "electro-hypnotizer" for $89.95 and Proteus, a supposed mind-control device that sold for $189.95. The site, which featured an atom with orbiting electrons, warned in big purple letters that some of their products might be "dangerous" and "illegal."

"Franz, aren't they the crazy people?" I asked.

"Yeah, they're the crazy people!" Franz said, jumping up. "They're crazy, but they're wonderful."

Franz was getting pretty excited, particularly as he explained how the Gayl Blaster could be hooked up to someone's back like a leaf blower. He had a patent, he said. He promised to send pictures.

Amy started to giggle.

"You built this in your garage?" she asked.

"Actually, I built it in my living room," he corrected.

Amy began to look a bit concerned.

"What would it do?" she asked. "Is it just really loud?"

He picked up the paper towel and held it in front of us.

"There's evidence from animals that those sorts of

frequencies are disturbing," Franz replied with a lopsided smile. "Those things can be felt throughout the body."

"Which does what?" Amy asked. "I mean, are we talking blood and stuff?"

Franz suddenly got cagey and shrugged his shoulders a bit. "Acoustic weapons are still all very anecdotal," he said.

"Well, not if you built one in your living room," I interrupted, also starting to laugh, imagining the Pentagon's most advanced nonlethal weapon being built and tested in somebody's den. "What did you test it on, your cat or something?"

"On myself," Franz replied, shrugging his shoulders.

"You demonstrated it on yourself?" I said, beginning to wonder about Franz's overall state of mind.

"Yes, but never loud enough . . ." He paused, cleared his throat and looked around the quiet kitchen a bit nervously. "I had some pretty high DB [decibel] values, 120 DB. And I tested it in an anechoic chamber. But what I really wanted to do was demonstrate the directionality."

The volume used was too low to hurt anyone, he continued. But the sounds it created were "some really insidious stuff."

"Franz, this isn't at Gitmo, is it?" I asked, thinking of reports of Muslim prisoners held in Cuba who were forced to listen to endless loops of Britney Spears's music.

"Um, no. It's up at Picatinny Arsenal in New Jersey," he said. "I used their money, too."

Amy and I exchanged furtive glances.

"Well, what did the Navy say?" I asked.

"This was where I had problems with staff at the Naval Postgraduate School," Franz said. "There's a rule. There's a certain equation to optics and acoustics alike: aperture versus wavelength determines the spread.

"I said, 'No, with my little technique here, I'm going to use my interference here, and I'm going to get cancellations.

I'm going to get an uncharacteristically narrow beam for the aperture size because my aperture is only this big,'" he said, gesturing with his hands.

He grabbed the paper towel, and began to draw waves, propagating out from the acoustic weapon. It sort of looked scientific, I thought, and I had no idea what he was talking about anyway. But even Franz admitted that there were doubters among the Navy scientists he consulted.

"They said, 'That's bullshit, that's not going to work, that's not scientific. Show us the equations.'"

But Franz couldn't show them the equations, because Franz, despite his title, was not actually a scientist. He had some engineering background, but beyond basic math, he didn't know the equations, and didn't much care. He just had a hunch it would work.

"Franz," I asked. "Then how did you know it would work?"

"It was in my head!" he answered, raising both his hands to his forehead. "I saw it visually. I saw this thing, you could do this, and turn it on end, concentrically, with speaker elements all around this thing carefully phased. You could get this beam out of the end of this thing. And the directionality of the beam would be uncharacteristic for that frequency, with that aperture size."

"But how did you know that without the equations?" I interrupted, pausing to look again at the paper towel.

"It just made sense," he replied.

"Scientific instinct?" Amy asked.

"I'm not going to qualify as a scientist," Franz replied modestly.

No, he just saw it. He saw the waves, the idea, the weapon. It was all in his head. The kitchen had suddenly grown very quiet as all three of us just looked down at the paper towel drawing of the Gayl Blaster. After all, where does one go from there? The coffee machine gurgled, the florescent lights hummed above us, and for a split second, I imagined the paper towel being confiscated someday by the Pentagon's security service, which would stamp it with a large black TOP SECRET. Amy, Franz, and I would all be sworn to secrecy, forbidden to talk about the military's now-classified acoustic weapon.

A malfunctioning fluorescent light flickered, and I snapped back to reality. Franz picked up the paper towel, paused to look at it for another moment, and then tossed it in the garbage.

Had Franz really built a weapon that flew in the face of scientific laws? Was the paper towel that had just been tossed in the garbage a schematic for a new class of weapons well advanced beyond what the military uses today? It was hard to say, but certainly Franz's assertion that his work had demonstrated capabilities that violated accepted physics was a bit dubious.

Amy excused herself to make a phone call, and I escorted Franz to the lobby as he continued to talk. There was another weapon he'd been working on that he wanted to tell me about, but he wasn't sure he could. This one was made from microwave devices. Not the high-tech microwaves built by military contractors, but the kind used to heat up frozen burritos.

This weapon would be used, he told me, on a particular part of the body. He hesitated for a moment and glanced around nervously, and then leaned in toward me, touching the back of his head slightly.

"You mean the brain?" I gasped. "What does it do?"

"I can't tell you that yet," he whispered. "Not now."

Then he smiled cheerfully and waved goodbye.

And with that, Franz Gayl, possible creator of the world's first acoustic weapon, walked away.

I stood transfixed at the front entrance as I watched his fading silhouette—slightly blurred from the heat coming off the downtown concrete—disappear off into the busy Washington street.

And for a second, I felt slightly disoriented.

This moment was what I have come to understand as the gateway to the Pentagon's scientific underworld. It starts the moment a military official tosses aside science and proceeds with some idea too far out to believe. Franz, who was often the first to admit his ideas were a little crazy, possessed a passion for exploration and curiosity.

I'm pretty sure that his desire to challenge the laws of physics would end if he found his weapons didn't work (several months later, I indeed uncovered a study sponsored by the Air Force that tested the Gayl Blaster on unsuspecting goats, who appeared rather unperturbed by the acoustic waves, and I have no reason to believe Franz has pursued this project). In other cases, however, I would learn that Pentagon officials enter the underworld of science, never again to leave. That's when things can go terribly, terribly wrong.

This book is about an idea that went terribly wrong.

It was on Halloween, October 31, 2003, that I first learned about the Pentagon's plans to build a frightening new bomb. Not just any bomb, but a weapon so powerful it could melt human flesh, penetrate deep underground into hardened bunkers, and evaporate whole city blocks with one fell blast. It would be unlike any bomb mankind had ever created, producing a powerful burst of gamma rays, delivering fatal doses of radiation in just seconds. It could create a powerful fireball, like a nuclear weapon, spewing radioactive material far and wide. In the hands of a suicide bomber, such a weapon could prove devastating, allowing terrorists to threaten entire cities with just a single explosive-laden car. I would later learn that the Pentagon believed the Russians had been working on their own version of the bomb, and so, too, perhaps, were the North Koreans. The military, in official documents, was calling this new bomb a revolution in warfare, comparable to the discovery of nuclear fission, and foreign journalists were issuing dire warnings that a new arms race was on the horizon.

It was a bomb so terrifying that one official from the Central Intelligence Agency had followed its worldwide progress for over a decade, and the State Department's arms control experts, concerned about the threat of a new era of proliferation, were calling up Pentagon officials, demanding explanations,.

And in the fall of 2003, when I first learned about this bomb, the Pentagon had just assembled the nation's top experts to look into mass-producing the critical material to be used in the new weapon. It was a material so precious and expensive, that a mere gram would cost as much as $1 billion. Building the facilities to produce the material could run in the tens of billions of dollars. A tidy sum, but worthwhile perhaps, if the alternative were to allow foreign countries to surpass the United States in military power. But most frightening of all, such a weapon, once built, might not fall under any existing arms control agreement. The system of checks and balances meant to contain weapons of mass destruction, however flawed and ineffective, would collapse into a meaningless pile of papers.

The world could be facing a new arms race of frightening proportions.

So how was it then, that the Pentagon had embarked on a quest for a new superbomb and no one in the United States seemed to notice?

Maybe it was because in the fall of 2003, the U.S. national security establishment was busy looking for nonexistent weapons of mass destruction in Iraq. Or maybe it was because Congress, with little public attention, was getting ready to roll back a decade-old ban that prohibited work on new nuclear weapons and low-yield nuclear weapons—so-called mini-nukes. Or maybe it was because this frightening new bomb never existed.

It was, in the purest sense, just an imaginary weapon.

As it was, I only heard about this weapon by chance, through a last-minute decision to attend a seminar on Capitol Hill, one of dozens of such briefings that take place in Washington every day. Stephen Younger, a senior Pentagon weapons scientist, was making the case for why the country needed new nuclear weapons. Off the record.

It was the typical crowd that shows up at talks in Washington on nuclear weapons—a mix of grim-faced policy wonks and aging social activists. A woman identifying herself as a psychologist raised her hand to ask a question, only to launch into a ten-minute diatribe against war and violence. Younger looked physically pained. Someone asked about so-called bunker-busting nukes, nuclear weapons designed to penetrate deep underground. Younger cheered up—he liked that topic.

And then a man behind me asked Younger what he thought of the hafnium bomb.

Younger, a nuclear weapons designer by training, just chuckled, as if the question was some inside joke that only a select few people would really appreciate.

After the talk finished, and people started shuffling for the door, I turned around to catch the man who had asked about the bomb. I couldn't help but notice that someone had passed him an "official use only" report, and he somehow possessed an air of authority.

I'd barely caught the name of the bomb, but I was certain I had never heard of such a weapon.

"What's a hafnium bomb?" I whispered.

He squinted at me for a moment, and then smiled mischievously. "Call me," he replied, handing me a card, and excusing himself quickly.

The plain white card read: Peter Zimmerman, Scientific Consultant.

I returned to my desk in the Senate press gallery, and for a few hours, I forgot about the hafnium bomb entirely, until I was about ready to leave work for the day and spotted the plain white card lying precariously close to the edge of my desk. I paused for a moment—thinking about all the things I had to do that week—and then I picked up the phone and called Zimmerman.

The next day, sitting over coffee in the bottom of the Russell Senate office building, Zimmerman began to explain the hafnium bomb.

Like many physicists, Zimmerman talked with his hands, sketching out diagrams, formulas, and equations as he spoke. He leaned over the table, grabbed my notebook, and began tracing out the shape of the second metastable state of hafnium-178. Oblong, spinning on an axis, it looked like an elongated football. Using stick-figure diagrams, he tried to show how this "charged up" variant of a regular atomic nucleus, called an isomer, would decay over time, emitting its energy in the form of gamma rays.

Doing what physicists call "back of the envelope" calculations, Zimmerman explained what made the obscure hafnium isomer so attractive to the Pentagon. It packed one hell of a wallop. One gram of the isomer material would have the same energy as one-third of a ton of TNT. The explosive force, if it could be released, would approach that of nuclear fission. I nodded my head, pretending to follow a bit of what he was describing. I didn't understand much of anything.

But then it got weird.

The Pentagon, according to Zimmerman, had started up a project to use this material to build a bomb far more powerful than any conventional explosive; the notional concept, he told me, was to make a nuclear hand grenade—a powerful device in an extremely small package. And the whole thing was a based on an experiment conducted in 1998 that involved a dental X-ray machine and a scientist from an obscure university in Texas.

"A dental X-ray," I asked?

"A second-hand dental X-ray," he corrected with a smile.

There was another issue. Prominent scientists had repeated the original experiment and found nothing—no results. And yet the Pentagon was starting down the path of what could eventually become a massive weapons project.

"So, it's like a nuclear weapon?" I asked, trying not to sound stupid.

"Yes," he replied, "but it'll never work." Worse yet, if it did work, it would in some ways be worse than a nuclear weapon because if it could be used to create an explosive force, it would disperse highly radioactive material far and wide. Hafnium-178 had a thirty-one-year half-life—the amount of time required for an element to lose half its energy—dwarfing the radioactivity of uranium, the material used in nuclear weapons. The isomer bomb would be the "the world's best radiological dispersal advice," he insisted.

But was he sure it wouldn't work?

Zimmerman paused for a moment, his eyes latched on to a table adjacent ours, where a Navy lieutenant holding a security pouch used to transport classified documents was chatting with a tall man in civilian clothing. Zimmerman shook his head in disapproval as the lieutenant nonchalantly unzipped the pouch, pulling out a piece of paper—the sort of casual violation of security procedures that happens everyday in Washington. Zimmerman's eyes narrowed for a moment in concentration, distracted by the security breach. But then he seemed to remember the hafnium bomb, and returned to my question.

"It's a bit like Potter Stewart's definition of pornography," he said, referring to the Supreme Court justice's famous 1973

"I know when I see it" declaration. "One knows bad science and after a while learns to recognize it as such," Zimmerman replied firmly. That said, he had checked out the claims.

Zimmerman explained that when he served in government a few years back, he had commissioned a high-level panel of elite scientists, a secretive group called the JASONs, to look at the hafnium bomb.

"The JASONs," I asked? I had been reporting on the military and the weapons business for several years, and had never heard of any group called the JASONs. It was as if there was a whole underworld of national security I knew nothing about.

The JASONs, Zimmerman explained, were a group of elite scientists who held some of the highest security clearances, access provided to them so that they could review any technical matter facing government, from North Korean proliferation to nuclear weapons testing. For hafnium, the JASONs' determination was conclusive: the science was dubious and such a weapon was simply not feasible. Hafnium couldn't be made into a bomb for a host of scientific and technical reasons: it was too radioactive to be around; it was too expensive to produce; and there was no way to get the chain reaction needed for an explosion. The original experiment, the panel concluded, was not solid proof of anything.

In essence, the JASONs concluded that the whole thing didn't pass the "snicker test," as Zimmerman summarized the assessment.

And other things about the initial experiment struck Zimmerman as bad science. The Texas scientist's former student was disavowing the results. And, he added, there were other rumors, though unconfirmed, that there had been some funny business with obtaining the hafnium. And of course, there was a basic issue of the experiment: a *dental* X-ray machine?

Despite all that, the Pentagon, according to Zimmerman,

has already assembled a group of senior scientists to look at ways to mass-produce the hafnium isomer, an endeavor that could run in the tens of billions of dollars. One of the scientists on that production panel, a Stanford physicist named William Herrmannsfeldt, was trying to kill the program, he told me.

"You should call him," he said.

For a moment, I considered the possibility Zimmerman was pulling my leg. I tried to imagine calling up a scientist I had never met, and asking about a secret Pentagon nuclear hand grenade. If it were true, I couldn't imagine he'd talk about it, and if it weren't true, I'd sound ridiculous. There was something about the whole idea that seemed unbelievable.

If the weapon couldn't work, I asked, what difference did it make what the Pentagon dreamt up, be it nuclear-style bazookas that could launch a city-destroying shell or a nuclear hand grenade? At some point, like with dozens of other harebrained military schemes, the Pentagon would simply give up.

Zimmerman, paused, leaned back in his chair, and then gave a more philosophical reply. He had a long-term view of the isomer bomb.

"I think that a program like this, once started, will have an enormous amount of inertia," he said. "It won't be a matter of someone deciding that this is just a waste of money, it will stick around for years, and likely grow in scope."

His concerns also went well beyond the science.

"Then there is the international relations aspect of it," he said. "Do we want to advertise that we are going to build a new nuclear-type weapon based on new physical principles? What sort of message does this send to our allies, to other countries?"

Dental X-rays, big bombs, and mad scientists. The whole thing sounded implausible. The Pentagon was certainly known for pursuing some unconventional ideas, but building a superbomb based on a dental X-ray machine?

But by the time we bid goodbye, I already knew that I would call William Herrmannsfeldt, and that I would try to learn more about this hafnium bomb. I had embarked on a journey that day, and without quite knowing it, I had crossed some threshold. I was entering a scientific underworld I never knew existed. This book is an attempt to explore that fuzzy line between military science and official foolishness—when an unusual little experiment becomes a Pentagon boondoggle.

The hafnium bomb began with the idea of building a handheld device with the explosive power approaching that of a nuclear bomb. It was in the truest sense of the word, an imaginary weapon. In the end, all it took was a used dental X-ray machine, a few die-hard supporters, some farfetched claims of a new arms race, and the Pentagon thought it was on its way to the next superbomb.

1

MICKEY MOUSE'S HAND GRENADE

IN THE SUMMER of 2002, Tony Tether, the head of the Pentagon's Defense Advanced Research Projects Agency was having a good day. He was at Disneyland, it was a sunny day in California, and the U.S. military was still basking in the glow of apparent victory in Afghanistan.

After ten years of declining defense budgets and a large dose of benign neglect, Tether's agency was flush with money and ideas. America post-9/11 needed its military again, and DARPA, the Pentagon's leading agency in charge of developing new military technology, was more important than ever. Its budget was up, its stature had grown, and many of the agency's inventions, such as unmanned spy planes, were flying high over Afghanistan.

"I thought there was nothing more appropriate than having DARPATech at Disneyland," Tether told the audience of in-house managers and entrepreneurs hoping to find funding for their ideas. "Disneyland is a land of dreams and fantasy becoming reality, and that is what DARPA does and does well."

Amid hulking statues of Mickey Mouse, the annual DARPA convention, DARPATech, featured advanced weapons and high-tech paraphernalia more *Star Trek* than Snow White, but Tether felt right at home. In fact, DARPA's new director was pretty confident at how things were going.

Money was pouring in and new ideas—*his* ideas—were being funded. The theme for 2002 was "transforming fantasy," and in a move that reflected Tether's creative streak, he had each of DARPA's offices build a display booth designed after a Disney theme—sort of an "It's a Small World" of weaponry.

DARPA conventions have always been a bit unusual among military trade shows. But since moving to Disneyland in 2002, the convention had taken an even more fanciful twist as life-size Cinderellas and the occasional dwarf passed by military mockups of *Starship Trooper*-style exoskeletons. The annual gathering of defense officials and scientists makes for some odd company, with men in military uniforms mixing awkwardly with West Coast engineers sporting ponytails and goatees.

Trade shows are a uniquely American tradition, and U.S. military conventions are perhaps even more so. It would be difficult to imagine Russian aerospace engineers donning Mickey Mouse ears while discussing the latest enhancements in ballistic missile technology. DARPA's staff, by contrast, is known for being a bit on the eccentric side and Tether encouraged the impression, particularly at the annual convention, where office directors dressed in Disney paraphernalia, and talks were interspersed with piped-in theme music from *Star Wars*.

But DARPA is special and its conventions, particularly in recent years, have mixed visions of science fiction with images of military power. By 2002, DARPA's annual conference had perfectly melded the far limits of science with the most outrageous military fantasies, overlaid it with a touch of Disney schlock, and packaged it for mass consumption as high-tech shock and awe.

On the extreme that year at Disneyland was a short movie clip innocuously titled "Brain-Machine Interface." A long-running DARPA effort to control lab animals remotely, the project involved implanting devices in the brains of rats. The goal, according to a scientist speaking at the convention, would be to tap the human brain without the need for any implants. And so at the home of Mickey Mouse—the most adored rodent of the animal kingdom—DARPA lauded the creation of "Roborat," a rat whose mind would be controlled by a machine.

"Who knows . . . if we can eavesdrop on the brain, maybe we can sort out deceit from honesty, truth from fiction," the scientist declared in an enthusiastic endorsement of mind control.

Of course, the DARPA official didn't mention that Roborat wasn't exactly a new idea. In the 1960s, the Central Intelligence Agency had its own wacky offshoot of mind control, though it favored cats over rats. Dubbed "Acoustic Kitty," the peculiar project involved trying to remotely control a feline to eavesdrop on conversations. Perhaps the agency should have selected a less free-spirited member of the animal kingdom, because on its first live test, acoustic kitty wandered away from its eavesdropping site—a park bench—and was run down by a car within minutes. Cats, it turns out, have a mind of their own.

Tether smiled a lot, perhaps because in the world of military weapons development, being the head of DARPA is one of the best jobs around. In an agency that prizes ingenuity above all else, failure can be a sign of risk-taking rather than a black mark, and DARPA's science and technology projects are shielded from the periodic oversight and cost audits that plague the Pentagon's more expensive weapons. The dozen or so slide presentations provided by the various program managers included the sort of over-the-top imagery common at the Pentagon in the age of PowerPoint, but with a twist that fit DARPA's image of itself as an "out of the box" agency. In one presentation, liquid metal hands à la *Terminator II* reached out to encircle a shiny planet earth, and in another, a smiling Mickey Mouse was a "sorcerer's apprentice" to DARPA. Creating technologies for the military can be *fun*, Mickey seemed to be saying.

For nearly fifty years, DARPA had served as the Defense Department's incubator for new technologies. The agency was founded the year after Russia launched Sputnik, beating the United States to outer space and putting American science to shame. The new agency was intended to bring the military into the space age by creating an environment where no idea was too far out. The novice agency had some early runs of bad luck in the world of Washington bureaucratic infighting. Six months after its creation, the government moved space programs out of DARPA's purview, temporarily leaving the fledgling agency without a mission. But it recovered and had some incredible successes like Arpanet, the precursor of the modern Internet.

DARPA's approach has often been to take on projects that seem impossible, and the term *DARPA hard*—now a part of the agency's legacy—refers to technology challenges considered too difficult for other parts of the military's research and development laboratories. In fact, Tether reminded his audience, it's not a sin to fail at DARPA.

"Why?" he asked. "Because no one remembers the failure. That allows us to try again and again without the baggage usually carried at an organization where people are there for careers. DARPA is *Groundhog Day*. We do things over and over and over again."

The same tolerance of failure, perhaps, was granted to DARPA's employees. Also presenting that year at DARPATech was John Poindexter, the disgraced Navy admiral who shot to fame in the 1980s when he lied to Congress about the Iran-Contra affair, which involved the United States selling arms to Iran in the hopes of freeing American hostages held in Beirut. Poindexter, a physicist, was enjoying a second career at DARPA as the head of the agency's Information Awareness Office. While Poindexter's notoriety may have made him slightly unique at DARPA, his enthusiasm for his project—data mining—placed him squarely within the agency's legacy.

Poindexter also fit the latest definition of DARPA program managers, the officials in charge of the various technology projects. According to DARPA's own description, its program managers were "free-wheeling zealots in pursuit of their goals."

There is no such thing as a lifelong DARPA bureaucrat. The agency's program managers serve for four years and are then forced out; a rule designed to prevent the entrenched interests and mediocre risk-avoidance that plagues other parts of the federal government. DARPA may be dominated by military engineers—a cautious breed by nature—yet the agency is designed to push them to take risks, and more importantly, to risk failure.

But in more recent years, DARPA has faced stiff criticism from Washington critics who claim the agency is too focused on the short-term and has forgotten its legacy of scientific innovations that yield revolutionary advances. Tether, a Stanford-trained engineer with a penchant for slick PowerPoint briefings, was charged to bring back the old DARPA, the one that took risks. "High risk, high payoff" was his motto.

With his expansive waist, oversized glasses and slick, graying comb-over, Tether was the very portrait of the Pentagon bureaucrat, who comfortably walks through the revolving door that connects the Department of Defense to the defense industry. Tether earned his PhD in electrical engineering, spending the greater part of his career working either directly or indirectly for the Pentagon. Tether took over the leadership of DARPA just four months prior to the terrorist attacks of September 11. It was fortuitous timing.

The 9/11 attacks on New York and Washington became, in a sense, the ultimate test case for DARPA's existence. The United States, in the words of President Bush, faced "a new type of enemy," and the Pentagon needed to adapt tech\nologies to this threat. The war in Afghanistan provided one of the best opportunities to take advantage of DARPA's innovations. Tether touted everything from the agency's work on behavioral analysis of possible terrorists to a new project designed to create robots controlled directly by the human brain. Citing DARPA-sponsored work at MIT, Tether lauded researchers who had already demonstrated the initial workings of mind-controlled robots.

"The genie is out of the bottle on this possibility, and the nation that first gains this capability will dominate," Tether said. Within a few decades, the military would field a new class of warriors, with the intellect of a human and the immortality of robots, and ready to "kick some butt," as DARPA's director put it.

"It's coming," he promised.

"DARPA is the only place you can have such a program, and talk about it!" Tether enthused. It was a "great time" to be in military research and development, he added.

He was right.

The Pentagon, under the leadership of Secretary of Defense Donald Rumsfeld, was on the lookout for "transformational" technologies—weapons and tactics that change the way the military fights. Rumsfeld's vision of transformation was the very hallmark of DARPA's mission. The Pentagon, for starters, wanted a host of new weapons and tools to target terrorists and rogue states. The military needed bombs that could blast deep into the earth to reach command bunkers hidden beneath tons of concrete, or penetrate mountains where countries like North Korea hide weapons of mass destruction. In other words, the military needed big bombs, possibly even nuclear bombs, to reach what in Pentagon parlance are called "hard-to-defeat and deeply buried targets."

DARPA was ready to step up to bat, and not just with weapons. The military could use a host of new technologies to defeat terrorism, including computer programs to track possible terrorists and futuristic technology that could help predict attacks. DARPA, Tether liked to remind people, was idea-driven, and any good idea that had a prayer of a chance of working would be considered for funding.

But by the end of 2003, with the occupation dragging out in Iraq, the mood in the Pentagon was very different. DARPA, moreover, was having a hard year—a very hard year. A couple of high-profile gaffes in DARPA's choice of "fantasy" had resulted in rare Congressional intervention in one case, and outright public humiliation in the other. First was the controversy that plagued Total Information Awareness—TIA for short—a data-mining system that would cross-reference personal databases to pinpoint the terrorists among us. Critics accused DARPA of funding a domestic spying system.

Then there was the Policy Analysis Market, popularly dubbed the "terrorism stock market," where private investors could bet on the probability of assassinations, hijackings, and general mayhem. The intriguing concept proposed by economists asserted that the private market could help predict political upheaval in the same way it spots economic trends. Investors would earn money based on accurate predictions. "Will terrorists attack Israel with bioweapons in the next year?" DARPA gave as an example to Congress.

Economists involved in the project defended the idea as sound, but DARPA wasn't winning points for good taste.

Congress, which rarely involves itself in the esoteric projects that DARPA pursues, was not amused. It certainly didn't help that Poindexter, the disgraced Iran-Contra figure, had headed up the office in charge of both programs. Poindexter, who had a PhD from the California Institute of Technology, may have seemed a logical choice to head a DARPA office, but he was a bad choice for public relations. A flurry of articles in the press accused DARPA of creating an Orwellian system to invade privacy, but Tether seemed vexed by all the negative publicity.

It wasn't enough, though, that Poindexter's Information Awareness Office ran programs with ominous sounding names like Genoa II and Babylon. Tether didn't seem to get what people saw when they looked at the office's symbol: an all-seeing eye placed in the center of the Masonic pyramid. It was ripe material for the op-ed pages of the *New York Times* and *Washington Post*. When the articles started coming out, first in a trickle, and later in a flood, Tether thought that the furor would eventually die down. TIA was, after all, just a computer program.

But with Congress threatening to shut down the entire Information Awareness Office, Poindexter was forced to resign. Later, DARPA quietly removed from its Web site Poindexter's presentation from the Disneyland conference, as if deleting his talk would make the whole fiasco go away. For an agency that was supposed to be on the cutting edge of data-mining technology, it seemed slightly odd that it didn't realize Poindexter's talk had been downloaded and cached on hundreds of servers.

In September of 2003, just a little over eighteen months after his speech at Disneyland, Tether sat over soggy eggs in front of a group of Washington defense writers, trying to explain what went wrong. The Defense Writers Group—a collection of a dozen or so defense and Pentagon reporters who meet regularly with senior government officials over breakfast at a hotel in Washington's Foggy Bottom neighborhood—had some controversial guests over the years, but Robert Dudney, organizer and leader of the roundtable discussions, always stuck with near-authoritarian control to the "one man, one question" rule. The system is typically executed with flawless precision—the reporter signals to Dudney, who acknowledges the unspoken gesture with a nod, adding the reporter's name to the list. When their turn comes up, they are allowed one question and a courtesy follow-up. The guest rarely finishes breakfast, but at least the reporters each get a question.

In a rare slip, Dudney lost control the day Tether came and reporters fired out question after question about TIA.

It seemed almost unfair. After all, the nascent data-mining effort was dead, and Tether had already taken his lumps in the press.

"Now, we screwed up," Tether finally said in exasperation. But the problem, he continued, wasn't DARPA, it was the media. "When those articles came out, you folks in the press really don't understand, well maybe you do, but the power that you really have is incredible. I've never really been personally involved, but the articles that were appearing were all quoting each other. . . . I am reading these things and I know what is going on, but start to think, 'Holy cow, maybe I don't know what is going on.' But we didn't respond. Where we screwed up is that we didn't respond."

While Tether recovered his composure and went on to tout DARPA's more successful endeavors, he didn't mention a small project that had been plucked from the Air Force's research facilities just a few months after the 2002 annual meeting at Disneyland. The new program, inauspiciously titled Stimulated Isomer Energy Release, SIER for short, was just getting off the ground with a series of grants, worth about $7 million, sprinkled out to national labs and small universities around the country. The program manager in charge of the research, Dr. Martin Stickley, had just transferred to the agency from London, where he worked at the U.S. Air Force's European Office of Research and Development. Stickley, like Tether, was a military engineer with a career that spanned decades of service in national security.

The new project was Stickley's brainchild, a product of research the Air Force had sponsored in Eastern Europe and Texas. Now in DARPA's offices in northern Virginia, Stickley had high hopes for the program, which involved releasing energy from a rare and highly radioactive material. The Air Force had funded the research over the years, ostensibly as a possible future energy source or to fuel aircraft, but the project hadn't really gotten that far into applications.

Engines, in either case, weren't a big priority for DARPA in 2002, which was focused on the war on terrorism. The agency wanted weapons.

The material Stickley was hawking to DARPA was made from a charged-up form of an atomic nucleus, called a nuclear isomer. This material wasn't just another way to build a more powerful bomb, but rather, an entryway to an entirely new class of weapons that could herald a revolution in warfare. Foreign countries ranging from Eastern Europe to Asia were already hard at work on this technology, Stickley warned.

The bomb Stickley was pushing was not your everyday weapon. The energy release from nuclear isomers could rival that of nuclear explosions, yet it required much less material than that used in conventional bombs. Moreover, this miniature bomb would be based on the explosive power of gamma rays—invisible electromagnetic radiation that penetrates more deeply than standard X-rays. The highly powerful radiation travels at the speed of light and can pass through most solid materials. In high doses, gamma rays mimic the effects of a ray-bomb in cheap movies, heating up solid materials until they explode. Gamma radiation even has a place in comic book mythology. It was a gamma-ray bomb that turned mild-mannered weapons scientist Bruce Banner into the Incredible Hulk.

In essence, isomers were a potent explosive energy that could be fit in small packages. And at high enough emissions, gamma rays could superheat air and create an explosive blast, similar to a small nuclear weapon. That's precisely what DARPA had in mind. By the time Stickley assembled a group of senior scientists to advise him on how to produce the hafnium isomer, he had encapsulated his vision into a startling picture of a small hand grenade complete with a pull-out ring, and a caption that read: **Miniature Bomb. Explosive Yield, 2 kilotons. Size, five-inch diameter.**

2

FROM ROMANIA WITH LOVE

JACK SARFATTI IS a theoretical physicist, cofounder of the "legendary" Physics Consciousness Research Group, and at least according to his own account, possibly also the Messiah. Urban legend has it that Sarfatti can usually be found lecturing at Café Trieste in San Francisco's North Beach, pontificating on time-travel techniques, weightless warp drive, and "torsion field space weapons." Telepathy, UFO technology, and the physics of propellantless propulsion are just a few of his self-anointed areas of expertise. Sarfatti claims close ties to the secretive world of military intelligence and has waged a war with Jimmy Wales, the founder of Wikipedia, the online encyclopedia, to ensure his role as a CIA consultant is listed on his biographical entry. Deeply concerned about issues of national security, Sarfatti often poses such weighty questions as: "Are time machines dangerous weapons of mass destruction in the wrong hands?"

But mixed in with the bizarre is an almost dizzying array of real physics and mathematics—complex and elegantly derived. Even detractors have called him uncommonly brilliant. His prolific writings, often sent out in massive group e-mails, are strewn with equations and occasionally delve into tangents on how aliens would use wormholes to steal away earthly abductees. He has also famously developed the idea of a "God phone" that could decode messages from the

Almighty, something that could come in handy if he indeed turned out to be the Messiah.

More importantly, Sarfatti is one of the leaders of the "new physics movement," a 1960s-era offshoot of conventional science that adopted its own unique blend of quantum physics and eastern mysticism, leading to dalliances in remote viewing (otherwise known as clairvoyance), an obsession with alien technology, and dreams of infinite energy and exotic death-ray weapons. Sarfatti's journey into this world, he often recounts, began with a phone call from a "conscious computer" aboard an alien spaceship to his home in Flatbush, New York. Sarfatti was just thirteen. According to his account of that fated call, he was told he was one of four hundred people selected for work on a yet-to-be-determined special project. From that moment, Sarfatti's fate was sealed, and he began his path to the scientific underworld.

To mainstream science, however, Sarfatti is a notorious crackpot. But if the father of a weapon is the man who claims to have first proposed the idea, then perhaps paternity of the isomer weapon belongs to Sarfatti, who began his career as a more conventional, and even quite promising nuclear physicist. He claims to have been the first, in 1963, to conceive of the "nuclear isomer laser," also known as a gamma-ray laser, or graser. A gamma-ray laser, were it possible, would be incredibly powerful; a quantum leap beyond the relatively low-power lasers being built in the 1960s.

After moving to Cornell University in 1964, Sarfatti says he discussed the idea with Hans Bethe, the Nobel Laureate who played a critical role in the Manhattan Project—the top-secret World War II-era project to develop the atomic bomb—and is often considered one of the greatest scientists of the twentieth century. But Bethe, a theoretical physicist, was not terribly optimistic about using isomers in a weapon. According to Sarfatti, Bethe discouraged him from pursuing the graser, telling him that the "cross-sections"—in other words, the probability a nuclear reaction could occur—would be far too small for nuclear isomers to be used in a laser. (Interestingly, that would be the same argument used by conventional physicists when DARPA would begin its hafnium bomb project forty years later.)

Since no one is around to contradict Sarfatti's claims (Bethe died in 2005 at the age of ninety-eight), Sarfatti's assertion to being the father of the isomer weapon stands uncontested, if perhaps slightly dubious. In either case, Sarfatti recognized the physics problem at hand and agreed with Bethe's logic, dropping the idea and moving on in later years to what he considered more promising endeavors.

"My current work is much more important—using dark energy for warp drive and wormhole construction," Sarfatti wrote to me from San Francisco, when I contacted him recently about his early isomer work.

Of course, Sarfatti's suggestion for an isomer weapon in 1963 also corresponded closely to the date when George Baldwin, a physicist at the General Electric Company, first proposed a gamma-ray laser. It was the 1960s and the golden era of science, when several corporations, like GE, ran prestigious laboratories that employed scientists to pursue basic research at company expense, even though such work might never lead to anything with a commercial interest. The gamma-ray laser fit into that setup fairly well: it was just a bright idea on the drawing board. Futuristic, but not immediately absurd. Baldwin, in either case, never claimed that he was the first to propose the idea. And in later years, it also was suggested that a Russian scientist by the name of Lev Rivlin proposed the gamma-ray laser even earlier.

Rivlin, who possesses an old-school Slavic charm, said he filed the patent application in Russia for a gamma-ray laser on January 10, 1961, but that scheme would have been completely unknown in the West. It was, however, cited in an authoritative Russian journal in 1963, and in later years by Baldwin. But even Rivlin modestly tipped his hat to a far better-known scientist.

"The principle physical admissibility of stimulated gamma-emission was certainly clear from 1905 after discovering the radiation emission laws by Einstein," Rivlin wrote me from Moscow, apologizing for his stilted English.

So, was Rivlin, an obscure physicist from Russia, the father of the gamma-ray laser? Or was it Baldwin, a corporate scientist who eventually went on to work in the U.S. nuclear weapons complex? Or was it Sarfatti, who rejected the idea to pursue work on a God phone? When it comes to weapons—particularly those that don't exist—paternity is a tough thing to prove, and as one scientist remarked, no one was ever asked to take a blood test. The truth is, lots of scientists were thinking about isomers in the 1960s, mainly because of the invention of the laser, shorthand for Light Amplification by Stimulated Emission of Radiation.

A laser beam is normally created by exciting the electrons in a lasing medium—be it a solid, liquid, or gas. The excited electrons shake off their energy in the form of photons of light. When enough of the electrons are excited, or "pumped," in the lexicon of lasers, they eventually release a cascade of coherent photons that create a laser beam. The trick of a regular laser is to pump up the electrons in the lasing medium; in a gamma-ray laser, by contrast, the idea is to use a material already in an excited state, like nuclear isomers. The tricky part is forcing the nuclear isomers to release photons in a cascade that will give off coherent wavelengths of light, as would be needed for a laser. And that was the difficult part, because no one knew quite how to force nuclear isomers to release their photons, let alone if those photons could be focused in a coherent laser beam.

Paul Robinson, an experimental physicist who would later go on to become the head of Sandia National Laboratories in New Mexico, fondly recalled his days as a graduate student in the mid-1960s. Everyone was excited because it was at the advent of lasers, he said, and that, in turn, also led to speculation that a gamma-ray laser could be the next big breakthrough. The short wavelength and high-power density of such a laser could have revolutionary applications in areas ranging from medicine to weapons. In the case of Robinson, he and some of his fellow students used an isomer of phosphorus-32 obtained from a nuclear accelerator to test out some of the ideas for a gamma-ray laser. They would rush the sample to the chemistry department, and use a Bremmstrahlung X-ray to "X-ray the living hell" out of the isomer; they would then try to measure whether the isomer had released, or *triggered*, any of its energy.

"It was a fairly inelegant process," Robinson admitted.

The experiments were hard to conduct and even harder to measure. With all the radioactivity generated by the sample, it was hard to tell whether they were triggering anything, or just measuring background radiation. That led to another problem. In seven attempts to trigger the isomer—that essential first step in creating a laser powered by nuclear isomers—they were only able to detect any accelerated energy release one time. They were never able to figure out why they couldn't replicate the results and eventually gave up, Robinson said.

And with the exception of these few novice efforts—and Sarfatti's offhand consideration—work on isomer weapons was not really pursued with any vigor in the United States. In the Eastern bloc, however, work on isomers proceeded with some enthusiasm, which would later become a near-obsession for U.S. intelligence officials tasked with following Soviet military technology. Of course, it isn't the job of intelligence officials, most of whom aren't scientists, to figure out if such technology was feasible, but only if the Russians were working on it. That made fears of a gamma-ray weapon—realistic or not—very important.

FOR ONE TEXAS scientist, the isomer bomb was inextricably linked not to Russia, but to Richard Nixon and his groundbreaking trip to Romania in 1969. It was the height of the Cold War, and Nixon was looking for a way to drive a wedge through the Soviet bloc. A year prior, Soviet tanks had rolled into Czechoslovakia, and Romania, under the despotic leadership of Nicolae Ceausescu, wanted to distance itself from the Soviet Union. A nascent relationship with the United States was a prime opportunity for Ceausescu to thumb his nose at Moscow.

Speaking in front of the United Nations on October 23, 1970, Nixon made a surprising announcement: the United States was going to expand its scientific cooperation programs to Eastern Europe. Nixon announced that he intended to increase funding for the U.S. National Science Foundation to work with Yugoslavia, which had pursued an independent line from Moscow since Josip Tito fell out with Joseph Stalin in 1949, and with Romania. These partnerships would serve as "important fruits of the revitalized political relationship we now enjoy with those two countries," Nixon said. The United States wanted to demonstrate that it was leading the world in prestigious areas like nuclear research and space exploration, and what better way to underscore this than by extending expertise to countries that were so clearly behind. While cooperative efforts with communist countries might prove difficult, basic science and research, Nixon told the United Nations, is "as unequivocal and uncompromising as the laws of nature."

Well, maybe not in Romania. Science in the small Eastern European country that sits along the Danube River was going through turbulent times in the 1970s. Elena Ceausescu, the erratic and vain wife of dictator Nicolas Ceausescu, fashioned herself a scientist, and in 1970, with the help of Romania's fearless leader and a forged doctorate in chemistry, she was appointed as the head of Romania's National Council for

Science and Technology. She was the nation's top scientist, and in her new position, it was all that much easier to attach her name to other people's work, adding to her already lengthy, and fictitious, résumé.

That same year, Carl Collins, an aspiring young professor at the University of Texas at Dallas, submitted the first—and at that point the only—proposal to the U.S. government to work with Romanian scientists. It was an unusual professional move for a professor of physics from Texas, but it was also an opportunity to take advantage of a pot of money that didn't require intense competition with peers from large laboratories. It was still the heyday of laser research, and nuclear physics was thriving in the United States. But in 1971, the U.S.-Romanian Project in Atomic and Plasma Physics was born with a grant to Carl Collins from the U.S. National Science Foundation. A second-tier university was suddenly wedded to a third-tier country in science, and for the next decade, Collins didn't have to worry about competing for grants against his First World counterparts.

When Collins landed that year in Bucharest, the communist country's capital, it was strange days for science in Romania. As the nation's science leader, Mrs. Ceausescu was traveling the world, attempting to extort honorary degrees, with some success, from universities in the United States and Great Britain. Obsessed with status, she figured more degrees would garner her more respect.

And Collins, a young professor of physics from Texas, was traveling to Romania to work with physicists Denisa and Ioan-Iovitz Popescu, a married couple with a strong interest in the emerging field of lasers, and a fascination with scientific fame that rivaled that of Elena Ceausescu. Over the next decade, as the field of lasers blossomed in Russia and the United States, the American-Romanian team worked on the fringe, convinced that despite their meager equipment, they would make breakthroughs that would beat out the mainstream scientific establishment. Collins quickly developed personal as well as professional ties to the country when he met Doina, a pretty and bright student of English, whose mother worked at the Romanian Academy of Sciences. The two began dating, albeit with local difficulties. Romania was a totalitarian state, and the police detained the American-Romanian couple, according to Collins's later account, nearly every night. They would be brought to the police station, a report would be written, and the pattern would repeat itself on the next date. Sometimes they would be picked up twice in one day, and have to inform the police that they had all ready been written up. It was just a formality, according to Collins, but it reflected the degree of suspicion for foreigners.

It didn't matter—the two were in love and married in 1973; Doina eventually returned with Collins to the United States.

The Popescus, Collins's professional partners, always had a sense of the grandiose. Denisa Popescu was convinced that she had been passed over for the Nobel Prize four times (although sometimes this figure was more modestly changed to two)—a claim that would make most reputable scientists cringe with embarrassment. The lists of Nobel nominees aren't released, and reputable scientists know better than to assert they've been named. That didn't stop Denisa Popescu from listing the alleged nominations on her résumé.

Of course, anyone in the world can nominate someone for the Nobel Prize—but it's only previous prizewinners and members of the Swedish Academy of Science that count. In the case of Denisa Popescu, it was Carl Collins nominating her, for work they did together during their cooperative program. While Denisa Popescu claims to have been nominated three other times, there is no evidence that anyone other than Collins did the nominating. In a 1995 nomination to the Swedish Academy, Collins noted that the prize-worthy work was done in partnership with him. "The actual discovery was made in my lab in Dallas as part of the cooperative program mentioned above," Collins wrote.

Claiming that their work in the 1970s had been passed over for Nobel Prizes because the experiments were not done in a major lab with top-notch equipment, Collins began to lay the foundation for what would be similar claims in many areas: inventions that were stolen, ideas that were passed over, or groundbreaking work that went unrecognized. Other institutions were being credited for work he did, because their experiments were better funded and used more sophisticated equipment, he believed. The Popescus' discovery of multiphoton spectroscopy, Collins claimed in one letter, was passed over for the Nobel only because another publication came out first with better measurements. Collins asserted the Popescus' paper was submitted first.

The Popescus' résumés, posted in much later years on a personal Web site, were an odd jumble of asserted claims to fame and possible delusions of scientific grandeur. Along with recounting nominations to the Nobel Prize, whole sections of their résumés were devoted to listing where Nobel Laureates had cited their articles. In a field where articles by practice reference previous works, the number of times a Nobel Prize winner cites a scientist's work was an unusual *idée fixe*, but one the Popescus relished.

Ioan-Iovitz Popescu, Denisa's husband, also had a sense of the dramatic. When Denisa Popescu died of heart failure in 2003—after many years of illness—her husband posted a lengthy letter on his personal Web site. Written in English, the letter was addressed to his wife's last treating physician and described the pitched battle Ioan-Iovitz Popescu waged with the hospital's surgeon, whom he calls by the *nom de guerre* De Dracula. The letter, a rambling, barely coherent history of his wife's final days, muses on the comparable scientific ranking of the surgeon versus that of the Popescus. At one point, Popescu describes handing copies of his and his wife's

résumés to the surgeon, as if to demonstrate the caliber of
the people the doctor was treating. Popescu then accuses the
surgeon of killing his wife out of professional jealousy. The
surgeon, it turns out, had a rather low ranking on the Popescu
"impact factors" scale (Ioan-Iovitz had devised an "impact
factors" score as a way to measure scientific influence).

"Are those damned 'impact factors' the reason for killing
Denisa?" Popescu wrote on his Web site. "I can assure you that
I have made a lot of academic enemies, some of them known
but most of them unknown, therefore even more dangerous.
Is it possible that De Dracula is one of them . . . ?"

Yet the Romanian collaboration, at least in Collins's
eyes, was a grand success. Other scientists involved in these
exchanges were restricted to their hotels, often unable even
to meet with their partners, while Collins enjoyed full access,
he later recalled. But the real watershed came in 1978, when a
young Romanian graduate student mentioned to Collins the
concept of building a gamma-ray laser using nuclear isomers.
If such a laser could be built, it would be the sort of thing
that would earn its inventor a ticket to Stockholm to collect
the Nobel, precisely the sort of recognition that Collins and
his Romanian colleagues had been pining for.

Collins was hooked. And from 1978 until the early 1980s,
the Popescus and Collins moved out of the laser spectroscopy
business and on to the gamma-ray laser.

But there were also problems. By 1982, the official phase
of the National Science Foundation's cooperative program
with Romania was drawing to a quick and ignominious close.
Relations between Romania and the United States soured as
the Ceausescu regime swerved even further into paranoia
and repression. The National Science Foundation ended the
cooperative ventures, and with it Collins's grant.

Back at the lab in Texas—and with no funding in sight—
Collins returned to teaching and publishing occasional

articles on optics. But the dream of a revolutionary laser remained. In 1982, Collins sketched out the conceptual design for a gamma-ray laser. All he needed was money. Conveniently, the chill in relations between East and West opened up a new opportunity: the Reagan era and the biggest military buildup of the Cold War.

3

THE SECRET LIFE OF
THE ISOMER WEAPON

WHERE EXACTLY DID the idea of a nuclear isomer bomb originate? In a 1984 Arpanet discussion group called "Arms Discussion Digest," one scientist discussed the isomer as a possible explosive that could be used for "neutron bomb replacement," as a trigger "for fission-free fusion bombs, or as ultra-clean bombs for peaceful purposes."

Last on the list of applications was a small nuclear device.

Specifically listed? A "nuclear hand grenade."

In 2005, I was sitting in a windowless conference room at Los Alamos National Laboratory, conducting an interview with James Mercer-Smith, a nuclear weapons designer, when toward the end of the hour, our conversation slipped over to isomer weapons.

The designer laughed, recalling earlier work at the lab on isomer weapons. "It was Charlie Bowman's idea," Mercer-Smith said. Even I had heard of Charles Bowman, a Los Alamos scientist who had become the subject of national controversy in the mid-1990s when he asserted that nuclear waste stored at Yucca Mountain in Colorado would create a spontaneous nuclear explosion—an idea that was resoundingly criticized in the scientific community.

"He wanted to build an isomer bomb?" I asked.

"No, it was another crazy idea," Mercer-Smith said. "He wanted to make a gamma-ray laser."

It would never work, he continued. "It's not just energy; you have to get it out in particular forms."

That's hard to do, he added.

"The problem with all the isomers is that by the time you finish designing the device, what you've created is a very interesting nuclear weapon," he said. "You just throw away the isomer."

That argument apparently didn't stop Bowman, who patented a "neutron-driven gamma-ray laser" in 1989.

It was never built.

Isomers in many ways always belonged to a sort of scientific underworld, and it should really be no surprise that a hotbed of gamma-ray laser enthusiasts resided at Los Alamos, birthplace of the atomic bomb. But even there, colleagues scoffed at their work and in the United States, isomer research in the 1980s continued at a low level, confined largely to a small band of believers, a point recounted perhaps somewhat wistfully even in scientific articles on the subject.

"For approximately 20 [years] a small but dedicated group of researchers has been exploring the possibility of developing a gamma-ray laser (graser)," began one scholarly article written on the subject in the 1980s. Perhaps because the problem was so hard, the researchers working on it were that much more dedicated. At Los Alamos, researchers like Johndale Solem was one of the small band of scientists to insist that such a device might in fact be feasible.

"Scientists have been toying with the idea of a graser since the first lasers were demonstrated in 1961," Solem wrote in a 1979 Los Alamos report. "The notion has been carried through several cycles of popularity and contempt and, until recently, has been continually clouded by a morass of misunderstanding and misconception."

Solem's latest thoughts on creating a gamma-ray laser—described as "The Gimmick" in his report (a term borrowed from the early days of the Manhattan Project, when the scientists weren't allowed to say the word "bomb")—was to use the "brute force" of a nuclear explosion. All he needed to test the idea was a nuclear bomb and the isomer, tantulum-180 dissolved in lithium-7. There was just one problem, Solem acknowledged: the price for a sample of 5.1 percent tantulum-180 was $1,175 per milligram, a hefty sum. And then there was also the cost of setting off a nuclear bomb at the Nevada Test Site. That wasn't cheap either, he admitted.

While the costs might be high, "the size of the payoff (huge) and the probability of success (moderate) justify a substantial investment," Solem argued. Proof that a gamma-ray laser works would be revolutionary, and would "stimulate worldwide interest" in the technology. And while a massive new weapon wasn't likely to earn anyone the Nobel Prize (or "a ticket to Stockholm" as Solem put it) it would be a great accomplishment for Los Alamos.

A powerful gamma-ray laser would offer a variety of nifty applications: it could be used to transmit power from earth to space or used in medicine for holography or radiotherapy. And of course, there was one other attractive feature: "It would make one helluva weapon," Solem noted.

And after all, Los Alamos was a nuclear weapons lab. What sort of weapon would a gamma-ray device be? Gamma rays are the most energetic wavelength on the electromagnetic spectrum, which makes them deeply penetrating, able to go through steel and concrete. A gamma-ray weapon would promise an additional advantage, Solem argued. "Gamma radiation is also particularly lethal to personnel, the weakest link of most military systems," Solem wrote.

In other words, gamma rays kill people much better than old-fashioned high explosives.

Solem offered one final compelling reason to pursue the gamma-ray laser: the threat of an equivalent Russian weapon. "Frankly, it makes my skin crawl to see so much hard evidence in the open literature that the Soviets are carrying on an intensive research program in this area, and to realize that we are doing nothing," he wrote.

Despite Solem's dire warning of the Russia threat, the isomer weapon was not exactly given high priority in the U.S. national security establishment, at least not then. The Department of Energy, as the caretaker for nuclear weapons research, funded modest levels of work—in the hundreds of thousands of dollars—for researchers like George Baldwin, the onetime GE scientist who has since moved to work at Los Alamos and was pursuing work on the graser. But it was basic laboratory research, and far from being a major weapons project.

The situation, however, was about to change.

"MY FELLOW AMERICANS, thank you for sharing your time with me tonight." These were the opening words of Ronald Reagan's televised speech to the American people, announcing his plans to deploy a space-based missile shield that would protect the United States against a Soviet nuclear attack.

It was futuristic. It was challenging. It was incredibly unlikely. It was March 23, 1983, and the Strategic Defense Initiative—soon dubbed "Star Wars" by its critics—was born.

It was a seminal moment in late–Cold War history. Star Wars emerged from a meeting between Reagan and Edward Teller, then the head of the Lawrence Livermore National Laboratory in California. Reagan was an actor-turned-politician with a gift for communication; Teller, father of the hydrogen bomb, was a renowned physicist with dark visions of Soviet domination. The two men found common cause in the belief that weapons in space could defeat the Soviet threat. Teller, one of the best known "hawks" on defense issues, had long been at the center of controversy and was perhaps best known for pushing President Harry Truman to go forward with developing the hydrogen bomb, known as the superbomb, or just the "Super," for short. He is also still remembered for his devastating 1954 testimony in Washington, D.C., against Manhattan Project director J. Robert Oppenheimer ("I would like to see the vital interests of this country in hands which I understand better, and therefore trust more," Teller said).

Several decades later, the father of the Super had the ear of Reagan, and foremost on Teller's agenda was his vision of a space-based X-ray laser, pumped by a hydrogen bomb, that would shoot down intercontinental ballistic missiles. In other words, the United States was going to detonate thermonuclear weapons in space in order to take down thermonuclear weapons. Reagan liked the idea.

Reagan's investment in Star Wars provided an unexpected boost for the band of scientists who had been laboring away in obscurity on the gamma-ray laser. While the nuclear-pumped X-ray laser was the real "star" of Star Wars, the infusion of money into defense science and technology had some other knock-on effects. For starters, the gamma-ray laser, which for two decades had been more a scientific curiosity, went from a theoretical issue to one that the government was prepared to finally spend some serious money on.

With the burgeoning interest in the field of gamma-ray lasers, Collins had started traveling to conferences to pitch his idea. It was a stroke of incredible luck. Collins submitted his proposal for a gamma-ray laser based on nuclear isomers to the Pentagon, and despite less-than-stellar peer reviews, by 1985, Collins achieved something that far exceeded his wildest dreams: a military contract worth over $10 million. It would have been an astounding sum for a top-tier scientist. For an unknown physicist from the Lone Star state, it was nearly unprecedented.

Gerold Yonas, a longtime Sandia scientist, recalled with amusement how Collins first visited his office to present his fantastic isomers. Yonas, who was the deputy director of the office in charge of the Strategic Defense Initiative, insisted he had no role in giving money to Collins. He found the Texas scientist less than convincing. Despite Yonas's lack of enthusiasm, Collins ended up getting about half of the total budget for the gamma-ray laser project.

Was it all just good timing? Or did it help to be from Texas? Even in 2003, a picture providing a clue to Collins's windfall still hung in his lab: Phil Gramm, the Republican senator from Texas, standing together with Collins at a podium. The military contract landed Collins on the front page of the *Dallas Morning News*. According to one scientist with a dislike for Collins, Gramm, then an influential senator, secured military funding for the Texas scientist by threatening to cut off all money for SDI's advanced research unless Collins's work was supported. It was the sort of congressional backroom dealing that went on all the time in Washington, though perhaps a bit more unusual for such largesse to benefit a single scientist.

After a few lean years, things were finally looking up again for Collins.

The problem was, no one seemed to think Collins's physics quite added up. Charles Rhodes, a physicist from the University of Illinois at Chicago, got up and attacked Collins's work in public. Rhodes, who had his own farfetched laser schemes, made Collins a regular object of attack at the missile defense meetings. Other scientists did so a bit more privately, although with equal vehemence. Cal Walker, a physics professor at Johns Hopkins University, recounted his run-ins with Carl Collins in the 1980s as an unforgettable experience.

"His stuff was so full of holes," Walker recalled. "It was crap."

According to Walker, Collins had gone to Pentagon generals with extraordinary claims of putting the power of the sun in something the size of a pen (in later years, Collins was inclined to use a golf ball, perhaps a better metaphor for

Pentagon brass). But if Collins's claims sounded ridiculous to scientists, they had traction with the military because Collins was getting millions of dollars. "He got huge amounts of money. He's never done anything. There is no science with his name on it," Walker insisted.

The conflict between Walker and Collins had to do with a scientific phenomenon called *sidebands*—a spectroscopic phenomenon related to vibrations in iron foils immersed in radio frequency fields. In the 1970s, Walker had done an experiment that demonstrated the formation of sidebands in iron foils. It was an interesting—and initially unexpected—result.

"At first we were very excited, because it might be new physics," Walker recounted. Collins, working separately down in Texas, was also excited, and set off on his own experiments demonstrating sidebands.

It's always exciting for physicists when an experiment generates some new, unexpected result, but the question is whether the result is a breakthrough, or just some quirky effect of nature that lies well within the realm of conventional science. It's not unusual in science for someone to forward a possibly groundbreaking result, only to find out later that those results are nothing that new, or worse yet, wrong. The first thing a good scientist does, Walker said, is to see if there's some easy way to make the effect disappear under conditions that shouldn't make a difference. That's what Walker and his colleagues did; they fiddled with the experiment, even using vacuum grease on the foils, to see if the sidebands would go away. They did. Walker's team finally determined the sidebands were due to a phenomenon called *magnetostriction*—a fancy term for changes that take place in certain materials when subjected to a magnetic field—an effect similar to a bell resonating when struck by an object. There weren't any new exotic nuclear processes involved in the sidebands, according to Walker.

Down in Texas, however, Collins, working on his own

experiments, was convinced that the sidebands were due
to a nuclear process—something that, if true, might have
important implications for his work on nuclear isomers. If
these unexpected nuclear transitions were occurring, it might
mean that there was a way to trigger isomers into releasing
their energy. Collins would clamp the experimental material
between tissues and do experiments, Walker recalled. Other
times, he'd use cardboard. "He did weird stuff, and nobody
could reproduce it," Walker said.

And long after Walker had attributed his work to a conventional explanation, Collins kept doing his odd experiments
and trying to publish papers claiming the sidebands were
proof of a "multiphoton transition." Walker was astonished
at Collins's claims.

"He was a devotee of this crazy Russian theory on multiphoton transitions," Walker said.

Walker, who was retired from Johns Hopkins University
when I spoke to him in 2005, described Collins as a true believer.
"He's a poor enough physicist that he doesn't understand."

Walker's interactions with Collins took on a sort of
dismal pattern. Collins submitted scholarly papers, which
were handed to Walker for comment, and Walker in turn
would give negative reviews to Collins's work. Life went on
like that for a while, until in one final twist, Walker described
how Collins called him one day and offered to cut Walker
in on his military funding, if only he would stop trashing
his papers. Collins, an obscure physicist from Texas, was
getting millions in funding. Walker, a physicist at one of
the top universities in the United States, described himself
as being lucky at that time to get $150,000 a year in funding
for his experiments.

Walker declined.

But Collins's work on Star Wars meant that regardless of
what mainstream science thought of his work, he was able
to attend military conferences and continue his research

with substantial investment. The Strategic Defense Initiative Organization admitted that Collins's work got only mixed reviews, but said that wasn't so important. "Depending on where they are in a field that's this narrow, peers might look at it differently from the way we do," a spokesman for the agency told the *Dallas Morning News* of Collins's work. "We might have funded it even if the review had been unfavorable."

Scientists interviewed at the time, however, were as dismissive of Collins's work as Walker, the Hopkins physicist. George Baldwin, one of the early scientists to suggest schemes for a gamma-ray laser, offered an equally dismal appraisal of Collins's work.

"We frankly just don't believe the formulas they used in it [their analysis]," Baldwin told a reporter from the *Dallas Morning News* back in 1988. The SDI group, he continued, had "been sold quite a bill of goods." But at least some people were impressed with Collins's work. In 1989, the Texas Academy of Sciences named Collins "Scientist of the Year."

ONE OF THE government physicists involved in isomer research suggested that if I wanted a historical perspective on isomers, I should contact Mort Weiss, a scientist semi-retired from the Lawrence Livermore Laboratory in California. When I finally caught up with Weiss, it was just after the New Year in 2004. It was over forty years after the first proposal to build an isomer weapon. No such weapon existed, however. That didn't seem to bother Weiss, who was moving furniture around his home near Livermore, California. I was balancing a computer on a hotel bed in Steamboat Springs, Colorado.

"Do you understand what an isomer is?" was Weiss's first question when I called.

I thought of Peter Zimmerman's little diagram of the turned-over football. After that first lecture, I had rushed out to buy *The Idiot's Guide to Physics*. I was reading about what an isotope was, and still found that confusing, particularly when I realized that an isomer wasn't the same thing as an isotope, and the book didn't mention anything at all about isomers.

"Not exactly," I replied truthfully.

Weiss began to speak about nuclear isomers. The term, he explained, came from chemistry, and had to do with molecules that have the same weight, but with a different configuration. The term *isomer* was adopted by Igor Kurchatov, the Russian nuclear physicist famous for his work on building the Soviet Union's first atomic bomb, when scientists discovered that the nuclei of some isotopes can live in charged-up states. Like chemical isomers, nuclear isomers are the same atomic weight as the nuclei of regular isotopes, but differ only in the arrangement of their subparticles.

The word *isomer* was thus adopted by physicists to talk about things like hafnium-178m2, Weiss explained, a nucleus that is of the same atomic weight as the isotope hafnium-178, but structurally different. The nuclei of the regular hafnium isotope has 72 protons and 106 neutrons, but its isomer has a different arrangement of subparticles that allows it to exist in a charged-up state. And that's what made hafnium important. The hafnium isomer, for example, stores about two and half million electron volts of energy, or 2.5 MeV in scientific shorthand. By comparison, a single molecule of dynamite stores just one electron volt.

It is because of that energy storage that isomers tend to attract a lot of attention, Weiss explained. "I could write a hundred papers about nuclear physics and nobody except specialists would care, but if I write three lines about isomers my phone doesn't stop ringing." Particularly in the late 1980s, isomers were widely regarded as the hot topic in weapons research, and a lot of people thought it was going to be the next superbomb. Directed energy weapons were the new fashion in weapons research and isomers might lead to a gamma-ray laser—an extremely powerful directed energy weapon. There were even people outside the lab who were convinced that Weiss's work, which he considered to be very basic science, was a part of some secret isomer bomb project. People from three-letter agencies were showing up at the lab, asking for briefings on isomers.

One of the early isomer enthusiasts back in the day was Fred Ambrose, then working at the Defense Intelligence Agency. Ambrose showed up at the lab around 1987, requesting from Weiss a thorough debriefing on isomers. They went into a secure room at Livermore—complete with flashing blue lights—to discuss the threat of a Russian isomer weapon. At the time, Weiss simply told Ambrose it was too early to know if one could make a bomb from isomers. It didn't matter to Ambrose, who showed up again a few years later, this time with the Central Intelligence Agency, still convinced the Russians had an active isomer weapon effort underway. People like Ambrose, according to Weiss, were absolutely convinced that sooner or later, someone would make isomers into a weapon.

It was also around that time that Weiss first met Carl Collins, who had recently returned from his extended stay in Romania. Like Collins, Weiss had received funding from the Pentagon's Strategic Defense Initiative Organization to work on isomers as a possible source for a gamma-ray laser. Weiss was looking at something called shape isomers, while Collins was preoccupied with spin isomers. Spin isomers are nuclear isomers inhibited from losing their energy because they have a high angular momentum or "spin"; shape isomers are similarly inhibited by their unusual shape. And it was at a missile defense conference that Weiss first heard the Texas scientist speak. The conferences themselves were filled with colorful characters—summer warriors, as Weiss called them, a term he felt appropriate to describe "a group of opportunist contract-seekers who had nothing to give but their proposals."

The big money of the Star Wars days had a way of attracting the most opportunistic of scientists, but even for that group, Carl Collins was unique.

"You're going to hear Carl Collins speak—you're in for an experience," one colleague told Weiss. Collins's talks were like theater, Weiss was warned: half Baptist-preacher, half used-car salesman.

Lecturing on the topic of gamma-ray lasers, Collins lived up to his reputation.

"He was extremely charismatic," Weiss recalled.

"Physicists all look like they fell out of the dust bin," Weiss said, but Collins was elegantly dressed and gave a wonderful lecture. "And everyone thought it was nonsense."

"Why was it nonsense?" I asked.

Weiss sighed into the phone. Collins wanted to build a gamma-ray laser, but he was going about things in a very odd way. It seemed like he was taking random materials and zapping them with energy to see if they'd trigger, Weiss explained. A lot of people at the conference doubted his results and attacked his physics.

"The kindest said it wasn't even wrong," Weiss said. "That is not a term of endearment among physicists."

But the head of the project, a Navy official, asked if perhaps the Livermore scientists could take the Texas scientist under their wing, so to speak, and at least guide Collins down a path of good physics. At Weiss's invitation, Collins flew out to the Livermore lab in the late 1980s and spent the day with Weiss and John Anderson, the head of Livermore's physics division. It was a unique experience for the Livermore scientists, according to Weiss. Collins was polite and charming, and carefully picked out those elements of physics he liked, merrily discarding the rest.

"That is not the way science is done," Weiss noted.

As it was, research at Livermore, by the mid-1990s, was

making it clear to the scientists there that isomers of either variety, shape or spin, though intrinsically interesting and with some potential applications, weren't going to make for any sort of superweapon.

"If I had larceny in my heart, we might have spun out shape isomers for decades," Weiss said.

But he didn't, and Livermore's isomer program ended around 1995. The reprints of shape isomer studies went to a back wall of his office, and Weiss, who joked that he would have been the father of the shape isomer bomb—had it worked—thought that isomers, at least as a weapon, were a closed subject.

The Livermore work on an isomer weapon had ended over a decade ago, I noted, and yet in 2004, the Pentagon's research agency, DARPA, was again funding the idea: how long could it go on?

Weiss laughed.

"When do they quit? When Collins quits," he said.

"I think it'll go on," Weiss continued. "If there's a review and DARPA is appropriately embarrassed, maybe it'll end. In the absence of that I think they'll find a way to spin this out for a considerable time.

"It's a scandal."

EVEN IF HE wasn't exactly winning accolades from the scientific community, back in Texas, Collins was getting a lot of money. And in November of 1991, construction began on a linear accelerator at the small lab in Texas. The four-million-electron-volt accelerator, which required excavating a huge underground area, wasn't massive by some standards, but it was big for the lab in Texas. Collins would use the high-energy photons from the accelerator to try to trigger different isomeric materials. Brushing aside the objections of physicists who questioned the utility of zapping the different materials, Collins declared that he was just a short step away from his goal. The massive new accelerator was to be the final stage of Collins's quest to build his gamma-ray laser.

"All of the component concepts have now been demonstrated experimentally and the persistent but false tenets of theoretical dogma which have historically inhibited the development of a gamma-ray laser have been eliminated," Collins wrote in his 1992 report to the Pentagon.

Improvements in optics have made things feasible that weren't known in his original the blueprint he designed in 1982, Collins insisted. Collins also now claimed initial success in a paper he published in 1989 announcing that he had "triggered" tantulum-180, the only nuclear isomer that occurs naturally. There was just one problem. It took much more energy to trigger tantulum than the isomer released. In other words, it was completely useless for a gamma-ray laser, or for any other application. He would find other isomers, he claimed.

"There are no a priori obstacles to the realization of the gamma-ray laser," Collins wrote.

He just had to find the right material.

It was 1993, the multimillion-dollar linear accelerator was completed, and Collins was down to just three isotopes. He was so tantalizingly close, he thought, to finally realizing the gamma-ray laser. There was just one problem.

The Soviet Union was gone, the threat of a Russian missile attack suddenly remote, and under pressure to cut back military spending, then-president George Bush pulled the plug on the space-based elements of the missile defense shield.

The gamma-ray laser ended, too, and with it, Carl Collins's funding.

Over $6 million had been pumped into Collins's work. The massive underground accelerator no longer had any use. For the other scientists involved in the program, it was money and concrete poured down a big hole in Texas, as far as they were concerned. For Collins, it was end of the gamma-ray laser and the beginning of a search for a new sponsor.

For Weiss, the Livermore physicist, it meant he didn't have to go to the missile defense conferences anymore, and even more fortuitous, he wouldn't have to visit Collins's dusty Texas lab. Livermore hadn't built the isomer bomb, and Mort Weiss was only the father of the shape isomer bomb in the imagination of a few paranoid lunatics.

But years after the government pulled the plug on Star Wars, and over a decade after the Soviet Union collapsed, the pursuit of the isomer bomb would draw devotees and true believers across government, and most importantly, in the halls of the Pentagon.

As if a harbinger of things to come, the isomer bomb got its first public mention at a May 1997 conference dedicated to U.S. and Indian dialogue on nuclear weapons. An Indian official in attendance listed isomers as one of several new types of nuclear weapons. Frank von Hippel, a former senior government science official, caught off guard by the unusual suggestion, told the conference attendees: "I don't know what to say about isomer weapons. It seems very futuristic, too, but I don't know enough to explain why. I'll have to find out more about it."

John Shaner, a Los Alamos scientist in attendance, shot back with, "Sounds like a Livermore concept, doesn't it?" The audience broke out laughing.

4

DEEP IN THE HEART OF LOS ALAMOS

IN THE MID-1970S, in a remote part of New Mexico, a high-energy beam pounded away at a thick piece of metal. Year after year, the beam struck the target leaving behind a thick sludge of irradiated metal, creating the raw material for what would eventually become the hafnium bomb.

It was at the Los Alamos Meson Physics Facility in New Mexico, better known as LAMPF, and the beam stop was creating isotopes that could be used for medical diagnostics. LAMPF was a part of the Los Alamos National Laboratory, birthplace of the atomic bomb—an appropriate starting point for the would-be isomer bomb. Founded during the World War II-era Manhattan Project, Los Alamos brought together the nation's top scientists in the race to build the world's first nuclear weapon. Situated on an isolated mesa in northern New Mexico, the laboratory even to the present day maintains a peculiar mystique and a unique culture defined in part by that historic role in nuclear weapons.

LAMPF was a nuclear accelerator, and producing medical isotopes didn't really have anything at all to do with nuclear weapons. But the facility was built in the 1960s for curious physicists who wanted to perform nuclear research, and the ability to produce medical isotopes for cancer research was thrown in as a way to "sell" the facility—and expand the laboratory. And that was important, because even if you're in the business of nuclear weapons, a laboratory operates along the same principles as any company: growth is good.

Medical isotopes weren't the only things being created, however. Once in a while, the beam would have a weird interaction with the base metal, tantalum, knocking out a few protons and neutrons, leaving behind just a few specks of a rare radioactive material, hafnium-178m2, the nuclear isomer of hafnium-178.

In its ordinary state, hafnium is a silvery metal that is neither radioactive, nor particularly rare. The element was first discovered in Copenhagen in 1922 and made world famous when Niels Bohr announced its discovery that same year, upon receiving the Nobel Prize. Hungarian chemist George Charles de Hevesy, the codiscoverer of hafnium, was nominated over a dozen times for the Nobel Prize—most instances citing hafnium—before finally winning it twenty years later for yet other work. Today, hafnium is used in industrial processing and nuclear control rods. And while hafnium can be mined from the ground, its isomer must be produced: a slow, tedious, and extremely expensive process.

Hafnium-178m2—the "m2" referring to the second metastable state of hafnium—is a truly unusual material, to be sure. Most nuclear isomers have fleeting half-lives lasting just seconds, and others so brief that physicists can only theorize their existence. The first metastable state of hafnium has just a four-second half-life, for example. But hafnium-178m2, with its thirty-one-year radioactive half-life, is unusually stable and extremely radioactive.

A defining feature of this unusual isomer is what is called its *angular momentum,* a measure of the spin of its nucleus. The hafnium isomer's angular momentum is sixteen steps above the ground state, or to use one metaphor, is like an object perched atop a sixteen-step ladder. In that excited state—at the top of the ladder—it stores about 2.5 million electron volts, or 2.5 MeV—energy that slowly dissipates like a battery losing power while it sits on a shelf.

But no one in the 1970s needed, much less thought about, the rare isomer. And there it sat in the sludge of an accelerator at Los Alamos, unknown and unrecognized until 1981. That was the year Jerry Wilhelmy, a nuclear physicist from Los Alamos, gave a talk at a run-of-the-mill scientific conference in Denmark on the esoteric (and to military thinking, rather unsexy) topic "Nuclei Far from Stability Using Exotic Targets." Wilhelmy was only interested in learning more about the basic physics of isomers. In his presentation, Wilhelmy offhandedly mentioned the possibility of producing isomers at the Los Alamos accelerator, including hafnium-178.

Wilhelmy had no idea that around the same time he was giving his talk, a physicist from Texas with a penchant for cowboy hats and big ideas was working with colleagues in communist-era Romania, convinced that isomers could be something much more than an interesting science project. One day, that Texas scientist would take that nearly forgotten by-product of the Los Alamos lab, and turn it into the focus of a major Pentagon weapons program. But plans to build billion-dollar facilities to mass produce that very special material were still two decades away.

The next time that scientists took note of the obscure little isomer, it was in a talk that Carl Collins gave at the Sandia National Laboratories, in New Mexico. The Sandia lab, along with Lawrence Livermore National Laboratory in California and Los Alamos National Laboratory in New Mexico, is one of the three Department of Energy labs responsible for maintaining the nation's stockpile of nuclear weapons.

Los Alamos was founded in 1943 as the focal laboratory in the effort to build the world's first atomic bomb. Edward Teller, father of the Super, helped found Livermore just a few years later, and the two design labs engaged in a sometimes-fierce rivalry. As the "Z Division" of the Manhattan Project,

Sandia was assigned the engineering task of designing and fabricating the weapons, while Los Alamos was at the heart of physics work.

Sandia, an engineering facility, has always been the odd one out: It's located inside the Air Force's Kirtland Base in New Mexico; it has had stronger ties to the military; and it is often more closely involved with the nuts and bolts of advanced weapons research. And unlike Livermore and Los Alamos, which are run by the University of California, Lockheed Martin, the country's largest defense company, operates Sandia.

It was perhaps not surprising that, beginning in the 1990s, isomer-weapon enthusiasts began to cross paths with increasing frequency at Sandia. In 1993, Mort Weiss, the physicist from the Livermore lab, gave a presentation at Sandia on high-energy-density materials for an Air Force symposium on "advanced weapons concepts," devoted to drawing-board ideas for future weapons. Isomers, according to Weiss, offered the tantalizing possibility of creating a conventional bomb with near-nuclear energy levels.

The most attractive feature of isomers—a point that would be repeated many times over the next few years—was their potential explosive power. A single gram of isomeric material could release one hundred thousand times the equivalent energy of chemical explosives, Weiss told the scientists and military officials present that day. There was another potential advantage to isomers, Weiss noted: an "isomer-based explosive" would release its energy all as X-rays. In other words, unlike nuclear fusion (the process of smashing two atoms together) or fission (splitting atoms apart), isomers might offer a potentially "clean" energy source—that is, if the isomer wasn't too radioactive to begin with. An isomer weapon, at least in theory, wouldn't have the political and environmental ramifications of radioactivity.

There was one final political attraction for isomers:

studying bomb applications could possibly be done in a contained, above-ground facility, rather than through underground testing. This was a particularly important point. When Weiss gave his talk in 1993, the United States had ceased underground testing of nuclear weapons only a few months earlier. Resuming those tests, even for an isomer weapon, would be difficult—if not impossible—in the prevailing political climate. Weiss, however, wasn't arguing that isomer weapons were right around the corner, only that not enough was known about them, and they might have *theoretical* potential to become a weapon.

The true believers in the isomer weapon were already beginning to jump on the bandwagon. Weiss first ran into them in the late 1980s and early 1990s. One of them was Hill Roberts, a scientist from Alabama who showed up one day at the Livermore lab in 1991 to request a briefing on isomers. Roberts, it turns out, was well connected to military officials in Huntsville, a.k.a. Rocket City, the city where Wernher von Braun, a rocket scientist from Nazi Germany, had come to establish the U.S. space and missile program. Huntsville today is home to U.S. Army's Space and Missile Defense Command, and consequently, a lucrative location for many defense companies, like SRS Technologies, where Roberts worked. For Weiss, the encounter did not leave pleasant memories: He gave a briefing to Roberts, only to have the Alabama scientist turn around, with slightly modified slides, and announce that he intended to submit them to the military to get funding.

Roberts had removed any references to Livermore from the briefing slides and wanted to check with Weiss to ensure he hadn't made any mistakes. Moreover, the slides were marked "SRS proprietary"—Roberts was passing them off as a product of SRS Technologies. Livermore scientists were furious.

Perhaps realizing that a Livermore–SRS Technologies collaboration wasn't going to work, Roberts instead paired up with Carl Collins, who was making frequent visits to Sandia. In fact, it was around that same time, in the mid-1990s, that Collins first showed up with a couple of Russians at Kirtland Air Force Base. Collins met there with Patrick McDaniel, a nuclear physicist at the Air Force Research Laboratory, and proposed a hafnium "burn-up" experiment. McDaniel had no idea what Collins was talking about.

By *burn-up*, Collins and the Russians meant they wanted to stick a sample of hafnium under X-rays, and see if over time they could show that its radioactive decay was accelerated—proof that the X-rays had a "triggered" the isomer. McDaniel, who had done work on isomers for Fred Ambrose, the isomer enthusiast from the Defense Intelligence Agency, was certainly intrigued, and he turned to his friend Nancy Ries, a program manager at the Sandia lab. Ries not only found money to give to Carl Collins for isomer triggering and but also agreed to support an isomer conference in Predeal, Romania, scheduled for later that year.

Ries also was quickly won over by the charismatic Texan and his Eastern European acolytes. She grew convinced that hafnium triggering could be the scientific equivalent of Glenn Seaborg's discovery of plutonium, which paved the way for nuclear weapons. Meeting Collins, as she would later describe it to me, sparked "one of the most exciting physics journeys I have ever been involved in."

If that were true, then the 1995 conference in Romania would be seen as a seminal event. It was an unusual venue, perhaps. Predeal is a sleepy mountain resort in Transylvania; its main draw is Dracula's castle in nearby Bran. What exactly happened in 1995 in that Romanian tourist trap is clear: the event brought all the hafnium believers together. Ries, the Sandia lab official, supported the conference. Collins also invited Hill Roberts to present a paper, and Roberts, to return the favor, got his company, SRS Technologies, to pay for some of the costs. The rest of the attendees were a bevy of Eastern European scientists who specialized in isomer research.

In later years, Collins would cite the conference as the critical moment in his career, because at the close, participants agreed that a definitive experiment needed to be done to prove that isomer triggering worked and further concluded that hafnium-178m2 was the best isomer to use. Or, at least that was Collins's summary of the proceedings.

But perhaps equally significant were some of the extraordinary claims Roberts made at the conference. Gamma-ray lasers, nuclear batteries, and powerful weapons were just a few examples of the applications Roberts believed were possible. Another idea Roberts forwarded was to use the gamma-ray bursts from isomer weapons to destroy stores of chemical or biological materials—the deeply penetrating gamma rays would be particularly effective at zapping deadly viruses.

Addressing the conference-goers, Roberts made a dramatic pronouncement: isomer triggering, he said, would be the equal of the 1930s-era discovery of nuclear fission.

"The significance of the development of a gamma-ray laser could be just as awesome in effect as the initial nuclear experiment, if not in scientific scope."

THERE WAS ANOTHER fortuitous piece of timing for the proponents of isomer triggering: U.S. intelligence agencies were starting to take note. Unsurprisingly, they were prompted by fears sparked from the East. In the early 1990s, the Russians had also begun to ramp up their research programs in isomers. Yuri Oganessian, the director of the renowned Joint Institute Nuclear Research, a premier Russian scientific center located in Dubna, just outside Moscow, used his professional clout to gain access to the institute's expensive cyclotron to produce the hafnium isomer. The institute wasn't able to make as much of the hafnium isomer as Los

Alamos did, but what the Russians did eventually produce was about ten times as pure at the U.S. sample. The production of such a large, high-quality hafnium sample got the attention of the U.S. intelligence services. That led to several visits by the "three-letter agencies" to scientists at the U.S. national laboratories.

That theoretical possibility of a weapon was enough to raise red flags for intelligence officials like James Corey, an intelligence analyst at Sandia, who first grew concerned about a possible Russian isomer weapon in the 1990s. The lab in those days was holding a Windows on Science lecture series, inviting former Eastern bloc scientists to speak to their Western counterparts. One popular topic was the gamma-ray laser, although the Eastern European scientists rarely mentioned explosives. But why spend all that time and energy on gamma-ray lasers, Corey wondered? The real interest, he surmised, had to be a weapon.

In 2005, when I reached Corey, he had just retired from Sandia and was living in the Missouri Ozarks—trying, as he told me, to find his "hillbilly roots." He was still very intrigued by isomers, he said, as well as by antigravity technology—the type of thing that will get you laughed out of most scientific lecture halls. That's all right, Corey said, he was an intelligence analyst, not a scientist.

Corey recalled the flurry of activity centered around isomers at Sandia back in the 1990s, including presentations by Carl Collins and the Russians. But Corey was mystified by what he considered the critics' personal campaign against Collins and his colleagues.

"They would follow them around, and give counter-presentations," Corey said. "I just thought it was ridiculous."

That the critics might simply not like Collins's science was apparently an idea that Corey had not considered.

For Jerry Wilhelmy, the Los Alamos scientist who had first identified the hafnium-178 isomer produced at the lab's accelerator, things took another unexpected turn. In 1995, a Russian group at St. Petersburg University headed by a physicist named German Skorobogatov claimed proof of triggering, or "stimulated emission," for an isomer of tellurium. In other words, they were saying—similar to later claims for hafnium—that the isomer could readily absorb X-rays, causing the excited nuclei to promptly release their stored energy. The Russian group reported that they had triggered the isomer when they cooled the sample to 10 degrees Kelvin (a very chilly minus 441.4 degrees Fahrenheit). When Los Alamos scientists, including Wilhelmy, attempted to repeat the experiment, they found no such results.

That led to a series of exchanges followed with yet another group of Russian scientists who accused the St. Petersburg–based group of fraud. A Russian scientist by the name of A.V. Davydov soon wrote to another Los Alamos scientist in 1992 complaining about Skorobogatov's experiment, saying "the details of work were without doubts invented by authors who proved 'too incompetent' to even bother to invent something plausible." Davydov also reported that yet another Russian scientist who had allegedly worked on that experiment declared "publicly at one of his visits to our laboratory that he did not participate in these works of Skorobogatov, did not sign the papers and even did not know about including of his name in the author's list. So we have to deal with extremely ugly thing; I did not meet anything like that during all my life."

Neither had Wilhelmy.

Claims of fraud, theft of data, and unlikely experimental results characterized isomer research from almost the start, and Wilhelmy, who liked to describe his physics career as otherwise "rather conventional," was beginning to learn that isomers attracted an odd group of people.

The U.S. intelligence community, which already feared a Russian isomer weapon, now had concerns over this latest development: a nascent Franco-Russian collaboration to look at triggering hafnium. Chantal Briançon, a French scientist, launched a partnership with Russian scientists from the Joint Institute of Nuclear Research. Using Russian-produced hafnium and French laboratory facilities, the two groups made plans for a joint experiment, and on March 24, 1996, in Orsay, France, the group bombarded hafnium isomer with alpha particles. For those in the U.S. intelligence community who trembled over the "isomer gap," the thought of a Franco-Russian hafnium bomb must have been that much that much more alarming.

The experiment, however, became the subject of much dispute. Collins, who claimed to be involved in the Franco-Russian experiment, immediately began to hype the results to Pentagon officials as definitive proof that hafnium could be triggered. There was just one major problem: the results from the Franco-Russian experiment were never published. And at least according to those involved in the work, Collins didn't have anything to do the collaboration.

"I received again questions about the results of our experiment in Orsay which have been reported by a Romanian fellow (I suppose C. Ur?)," Briançon later wrote to Wilhelmy, referring to Collins's student. "This seems to me very strange, as nobody from this company had access to the data and this person was never in the collaboration I am coordinating."

Briançon accused Collins's group of purloining her data, and in a more frank exchange with another U.S. scientist, she allegedly threatened to rip Collins to pieces; coming from Briançon, who is sometimes described as looking like a Soviet tractor, it seemed like a pretty realistic threat. That, however, didn't stop Collins from continuing to brief military officials on the French-Russian experiment as proof of isomer triggering, showing slides depicting Briançon's data. Collins also hinted that the "authorities" (presumably the French government) may have prevented the Orsay group from releasing that data, though he didn't identify who, exactly, those dark forces might be. But in later years, Collins and other hafnium supporters continued to push the idea that the French were pursuing a secret hafnium weapon with Russian help.

All this controversy left Collins with a major problem: with Briançon denying that the 1996 work proved isomer triggering, Collins needed to conduct his own experiment, and that meant getting his hands on his own supply of hafnium. The Russian isomer, at least according to Collins, had "disappeared" after the fateful 1996 experiment, and there was only one other supply of the hafnium isomer: Los Alamos National Laboratories in New Mexico. The only problem was money.

One of Collins's supporters came through. Roberts, it seems, had persuaded some government friends at the Pentagon's Ballistic Missile Defense Organization, the successor to the Star Wars–era Strategic Defense Initiative Organization, to give Collins money to buy the hafnium. According to Roberts, he was able to get some government people "who had some extra money" to help Collins buy a hafnium sample from Los Alamos. That funding followed a convoluted path, however.

The money, like the earlier Star Wars contract, was funneled to Collins through the U.S. Naval Research Laboratory in Washington, D.C. And Paul Kepple, a government scientist who had managed Collins's earlier contracts (and who, according to other scientists, would happily have never given a dime to Collins in the first place) was now tasked with providing yet another contract—albeit a modest one—to the Texas scientist. The contract was for about $100,000, Kepple later recalled, and was for one purpose: to buy hafnium from Los Alamos. But the most unusual part of the arrangement was the actual sale of the hafnium: it was purchased by Coherent

Technologies, a company registered at Collins's home address, with his wife, Doina, listed as the sole corporate officer.

Collins never mentioned that his wife's company owned the sample, but simply said that "networking" had allowed him "free access" to a hafnium sample. The U.S. Navy was not much better. It took several months, a half-dozen calls, and finally the threat of a Freedom of Information request before the Naval Research Laboratory finally agreed to provide information on that sale. A decade after the purchase, in 2005, when I asked the Naval Research Laboratory why they had given the contract to a private company, not to mention one owned by Collins's wife, Kepple, the Navy official in charge of the contract, replied with an unusual explanation: "I suspect that the reason is that the material obtained from Los Alamos was radioactive and there is much more paperwork for radioactive materials at the University than in a small company."

Intelligence agencies were worried about Russian isomer bombs, yet the Pentagon was giving money for a private company—registered at a private residence—to buy highly radioactive materials for a possible weapon because it required less paperwork. It would have been a highly unusual arrangement in most parts of government-sponsored science, but perhaps for isomers, it was business as usual.

Whatever the case, Collins now possessed a piece of hafnium, and he began to prepare for what he was sure would be the breakthrough experiment: triggering hafnium with nothing more than a small X-ray machine. Would it work? Just prior to his big experiment, he gave an interview to the *Dallas Morning News,* exuding confidence about what most scientists considered the unlikely prospect of triggering hafnium. Hafnium, Collins declared to the local press just days before the seminal experiment, was ready to be triggered.

"I wouldn't carry this sample through an X-ray machine at an airport."

5

THE DENTAL X-RAY GOES TO WAR

IT CAME FROM Los Alamos, express delivery. Refined, processed, and sealed in plastic, it looked more like a fine layer of dust than something that might cost as much as $28 billion an ounce. In a barn-like lab in Texas, among massive accelerators, old pieces of cannibalized metal, layers of dust, broken knobs, bits of wire, and discarded electronics, the precious material was placed atop an upside-down Styrofoam coffee cup.

A dental X-ray machine focused on the cup. Days passed. To the naked eye, nothing happened. But during that time an invisible X-ray beam, modulated by a commercial audio amplifier, slammed into the minute amount of material perched on top of the Styrofoam platform. Protected behind cinder blocks, a flickering computer screen registered jagged graphs, and a small group of men with radiation detectors attached to their shirts gathered around and watched and waited.

One Styrofoam cup, a few barely visible specks of hafnium-178, and a dentist's X-ray beaming away—it's hard to know exactly what happened that summer in the dusty lab in Texas, but in January 1999, the *Dallas Morning News*, under the section "atoms and the universe," gave Carl Collins's version of events: "Dallas scientists have succeeded in coaxing gamma ray emissions from a chemical sample by zapping it with X-rays from a dental machine."

The goal of what would be later dubbed "the dental X-ray experiment" was to prove what Carl Collins believed was possible based on his nearly two-decades long obsession: that nuclear isomers could be triggered with a relatively small dose of X-rays. Money was tight in the summer of 1998, but Collins had the sort of resourceful graduate students that only the former command economies of Eastern Europe could produce. For the X-ray source, a few of his students went out to a local Dallas business that specialized in refurbishing used dental equipment. With a little sweet-talking and fifteen hundred dollars, they were able to buy a used dental X-ray machine, the exact type used in thousands of strip malls and office parks all over the country. Another one of Collins's students had an even better idea: what if they hooked up a music amplifier to the dental X-ray machine, allowing them to modulate the energy output? They ended up buying a five-kilowatt amplifier that Collins later described to me as "the type used in rock concerts."

Working along with Collins on that experiment was a motley collection of scientists and students. There were four Romanian students, a French student, and one native Texan. There were also a few researchers from the former Soviet Union, including Sarkis Karamian from the Joint Institute for Nuclear Research in Dubna, Russia and Vladimir Kirischuk from the Institute for Nuclear Research in Kiev, Ukraine. Hill Roberts, the physicist who helped Collins get the hafnium, paid for the East European scientists to attend, hoping that his company would be in a prime position to profit from a major breakthrough.

Not everything went according to plan in the experiment, however. For starters, Collins's son, Carl Baxter Collins, an undergraduate at Texas A&M, was killed in a car accident in Romania in the middle of the experiment. For whatever reason—perhaps because of the time involved or the number of people who had flown to Texas to participate—

Collins chose to forge on. One can only speculate about what Collins's mental state might have been, conducting an experiment while mourning the death of his oldest son and namesake.

There was also a growing disagreement between Collins and his former student, James "Jeff" Carroll, who had taken a position at Youngstown State University in Ohio. Carroll's wife back in Ohio was pregnant, and Collins insisted that Carroll spend more time down in Texas. The experiment extended out for several weeks, but Carroll ended up attending for just a brief period, and by the end, he and Collins were barely on speaking terms.

Another festering dispute between the two men was over the financial relationship surrounding the hafnium experiment. Technically, what little government funding was provided for the experiment, by the Sandia lab, went through Carroll and his university. Yet Collins, whose wife's company owned the hafnium, was leasing it to Carroll for $63,000. So, in an oddly lucrative relationship, Collins was charging to use hafnium for an experiment at his own lab, with himself as the lead investigator. But for hafnium, that triangulated relationship was perhaps again simply business as usual.

Back at Livermore, Mort Weiss got word of the dental X-ray experiment when another reporter from the *Dallas Morning News* called him. The reporter announced that Collins had claimed to have "triggered" the hafnium with a dental X-ray machine, and wanted Weiss's opinion about the matter.

"It didn't violate laws of nature. But I thought it was very unlikely," was Weiss's response to the reporter.

But not long after that, Weiss got a call from a CIA consultant, again asking about the Collins results, which had been circulating around Washington as a prepublication paper. Since Weiss hadn't seen the paper, he called Collins himself to ask for a preprint. Collins declined, saying that he felt Weiss was "hostile," though he offered that if Weiss wanted to "join his team," he would cut him in on the program and share in the glory

Weiss declined.

When Weiss informed the CIA consultant that he wasn't able to see the paper, a copy was delivered promptly to him at Livermore the next day. It was at that point that Weiss saw that Collins's results really did violate physics. More precisely, on examining the paper, Weiss realized that Collins's results, if true, would violate the "sum rule," that is, well-established physics that determine how much energy a nucleus could release.

"I told him it was nonsense," Weiss recalled of his response to the consultant.

HOW A DENTAL X-ray experiment showed up in the world's premier physics publication is still a bit of mystery. Maybe the reviewers didn't read it that carefully, some scientists suggested. Perhaps they didn't bother to sit down and do the mathematical calculations that would have shown how the results violated the known laws of physics. Or perhaps the concept of peer review—the notion that other scientists should determine what papers are worthy of publication—doesn't always work as well as hoped. According to one version of events, the editor that month was out of town, and assigned the paper out to a less-than-competent reviewer.

Whatever the case, a paper entitled, "Accelerated Emission of Gamma Rays from the 31-yr Isomer of 178Hf Induced by X-ray Irradiation" showed up in the January 25, 1999 issue of *Physical Review Letters*, the leading physics journal. Worded in the subdued tones of scientific writing, the paper described the Texas group's experiment and concluded: "For the case of isomeric Hf-178 we have demonstrated the resonant absorption of an X-ray photon with the energy of the order of 40 [thousand electron volts] KeV can induce the prompt release of the 2.446 [million electron volts] MeV stored by the isomer into freely radiating states. This is an energy gain of about 60."

To someone who didn't know the science, the results might sound no different than any other indecipherable paper published in a physics journal. But what Collins was claiming was that he had done something truly miraculous: he had used relatively low-energy photons from the dental X-ray machine to trigger the release of a huge amount of energy stored in hafnium-178.

What exact did this mean and why did an obscure scientific result suddenly attract the interest of the military?

An isomer is a bit akin to a battery, as John Anderson, the former head of the Livermore lab's physics division, explained it to me. An isomer nucleus is "charged up," roughly similar to a battery. And like a battery that slowly drains its energy to run a small radio, an isomer will lose its energy through radioactive decay. Of course, with a radio, tapping the battery's energy is as simple as flicking a switch. An isomer, on the other hand, decays at a steady rate, which for hafnium is a thirty-one-year half-life, meaning it loses half its energy over thirty-one years. Nobody knows how to make a "switch" that taps that energy on demand. Trying to find a way to control that release, scientists have long believed, might lead to a useful energy source—a nuclear battery that could drive rockets through space, for example.

Of course, there has long been another possible application for isomers, which is to find a way to release that energy all at once, which would in theory make for a nice bomb, rather than a battery.

In the case of a weapon, Anderson told me, the isomer nucleus would be a bit more comparable to a blown-up balloon. A balloon will slowly leak air, or in the case of an isomer, lose its energy through natural decay. A gentle pinprick to the balloon, however, would effectively release all the balloon's energy at once, with a big bang. With an isomer, what scientists had long been looking for was a similar "pinprick" that could release the energy all at once, as in a bomb.

Collins's claim in 1999 was that low-energy photons from a dental X-ray had triggered the quick release of a large amount of energy from hafnium. If the results of Collins's experiment were as he reported, then he was well on the way—though not nearly all the way—to finding that gentle "pinprick" to release energy from the hafnium isomer, like popping a balloon.

Easier said than done. Among nuclear physicists, those results were met with some curiosity, some doubt, and a great deal of ridicule. The results were at odds with how nuclear physicists believed that isomers should behave and the energy output was well beyond what they believed was possible. Nuclear physicists took aim at a number of issues, focusing not just on the data, but how Collins chose to interpret it. In three negative comments published later in *Physical Review Letters*, critics attacked his science, challenged the statistical accuracy, and questioned the high margin of error he reported; the error bars were almost as large as the results. They also challenged the basic physics: Collins's results were ten thousand times greater than what conventional physics would predict.

The word *trigger* is what many scientists really took issue with. The question was not whether isomers could release energy, but whether it could be triggered using the low-energy X-rays that Collins described in his paper. Collins answered the negative comments with a sharply worded rebuttal, attacking the "judgmental opinion" and "logical fallacy" of the critics.

Collins didn't really answer the criticisms, but simply promised that "confirming details" would soon be available.

IF THE HAFNIUM bomb had an intellectual predecessor, it would have been cold fusion, the widely derided claim that nearly unlimited energy could be produced with little more than water. In 1989, the Cold War was drawing to a close, and many believed the biggest threat to national security was no longer the Soviet nuclear arsenal, but the world's dependence on a limited supply of fossil fuels. Nothing drove that point home harder than a massive oil spill off the coast of Alaska, an environmental catastrophe of epic proportions. The day the Exxon Valdez, guided by a drunken captain, created one of the largest environmental disasters in recent history, cold fusion—a possible solution to the world's energy crisis—hit the front pages of newspapers.

The basic history has become a caricature of bad science and is usually recounted in this abbreviated form: two chemists in Utah announced at a press conference on March 23, 1989 that they had tamed the power of the sun in a test tube, providing a potentially unlimited supply of clean energy. Their experiment involved a car battery, a few pieces of Tupperware, and driving an electric current through heavy water—water whose "light" hydrogen atoms have been replaced with "heavy" deuterium atoms.

While the experiment seemed crude, the two men were well-respected chemists, which gave their claims at least an aura of credibility. But the idea struck most nuclear physicists as bizarre; cold fusion defied the known laws of physics and the experiment itself looked amateurish. Nuclear fusion usually takes place in the sun or in thermonuclear weapons, where intense heat and pressure force nuclei to overcome their natural repulsion and fuse together. So it seemed that much stranger to most physicists that it didn't occur to the two chemists in Utah that their claims of fusion, if true, would have also unleashed a torrent of radiation, killing them on the spot. More importantly, the would-be discoverers of cold fusion broke convention to announce their results at a press conference, rather than through scientific publication.

When many laboratories were unable to replicate the results, the scientific community lined up against cold fusion and in less than a year, the "scientific discovery of the century" was relegated to the scrap heap of pseudoscience. That wasn't quite the end of the story, however. Over the next decade, cold fusion developed into a true underworld as scientists who were able, they believed, to reproduce the original experiment, refused to give up. A counterculture emerged, with scientists involved in the cold fusion field holding their own conferences and self-publishing their papers, all in a sort of parallel universe to mainstream science.

In 2004, I found myself in a Rockville, Maryland diner, sitting across from a member of the scientific underworld and one of the most famous cold fusion scientists: Peter Hagelstein, an MIT professor, a former rising star of physics, and at least according to popular accounts, a onetime protégé of Edward Teller. Hagelstein certainly wasn't interested in hafnium, but there was at least one important link between cold fusion and the isomer bomb: the Star Wars X-ray laser—that improbable futuristic weapon that would be pumped by a massive nuclear blast to wipe out an incoming Soviet missile strike. Hagelstein, an applied physicist, had been the father of the nuclear-pumped X-ray laser, the weapon that was going to be at the heart of the Reagan-era Star Wars shield. Enthusiasm for the X-ray laser had led to funding for the isomer-based gamma-ray laser.

Hagelstein, a self-described pacifist, left the military world at the height of Star Wars, and the X-ray laser was long since cancelled. For fifteen years, Hagelstein had worked on cold fusion, a science that was ridiculed by his colleagues, and seemed to attract mostly scientists on the margins of credibility. His reputation in tatters, his research funds long gone, Hagelstein was still doggedly pursuing cold fusion.

Now, at fifty years old, the scientist who had once been described by the press as "brilliant" and a "genius," worked almost exclusively on cold fusion. I wanted to know why. Why had Hagelstein dedicated himself to fringe science? I found myself having a difficult time making sense of it, and Hagelstein wasn't offering many clues.

Hagelstein took a sip of his lemonade and stared at me silently.

Our conversation began with cold fusion, but drifted over to the X-ray laser. Hagelstein's work on Star Wars gave him near-mythic status as the naïve scientist drawn in by the lure of high-technology weapons. Hagelstein himself encouraged that myth, and sitting in the diner that day, he described his work on a weapon as a "Faustian bargain" to get support for laboratory lasers he thought would benefit medicine.

"I didn't want to be remembered for having invented the X-ray laser, something pumped by a hydrogen bomb," he finally said.

"But, Peter, the bomb-driven X-ray laser doesn't exist," I pointed out, noting that the it was cancelled in 1992 amid myriad technical problems.

"It could if we started testing nuclear weapons again," he corrected with more than a touch of annoyance, which I thought was an odd reaction for someone who didn't like bombs.

It was an interesting academic question: could the X-ray laser exist? Critics argued the whole concept was preposterous, while the proponents claimed it was sound and just needed more time and money. Since most of the results from the early tests are still classified, it's hard to say, with any great precision, which side is right.

The X-ray laser was certainly a topic of debate, but Hagelstein's attachment to cold fusion was much harder to understand. His only explanation was that he believed the experimental results. And therein was the problem, because fifteen years later, reproducibility remained the most nagging issue for cold fusion. While proponents were claiming better results, Hagelstein acknowledged the experiments were far from perfect. Some results were never successfully replicated, he conceded. But reproducibility should not be the sole standard for the existence of a phenomenon, he told me. "Have you ever seen an earthquake?" asked Hagelstein, who grew up in California's San Fernando Valley. "Either something happens or it doesn't," he said.

I thought over Hagelstein's argument for a moment.

"When an earthquake takes place, we know it," I said, jabbing my fork into overcooked pancakes.

"How do you know if it's not reproducible?" he asked. "Suppose you imagined it? Maybe you're deluded. If I can't make it happen in my laboratory, I don't believe in it? The scientific question is one of whether it happened at all."

It was the sort of argument that couldn't be won. I never asked Hagelstein the question I meant to ask: if you don't have reproducibility, what do you have in the end? How do you ensure you aren't forever chasing your tail?

Finished with his lemonade, Hagelstein handed a dollar to a waiter collecting for muscular dystrophy, and gratefully accepted a ride to Reagan National Airport to pick up a cold-fusion colleague. Driving down the George Washington Parkway, Hagelstein narrated the twisted progress of fifteen years' work on cold fusion. As new experiments emerged, so, too, did his theories. For every strange, new result, Hagelstein had a new theory for how it worked. But Hagelstein had tossed aside as many theories as there were different experiments. As I dropped him off at the airport, I realized I still had no idea if Hagelstein was completely crazy or incredibly sane. He was convinced cold fusion was right, just as he was convinced the X-ray laser would have worked, and that was about as far as I was going to get to understanding his devotion to fringe science.

"Why do a certain set of scientists end up chasing a certain set of nearly hopeless scientific pursuits?" I asked Philip Coyle, a former associate director of the Livermore lab, who knew Hagelstein from the days of the X-ray laser. Coyle thought it over for a few days, before replying. "It's corny to say, but before long *The Man of la Mancha*, 'To dream the impossible dream . . .', and 'to follow that star no matter how hopeless . . .' all came to mind," he wrote me.

"But, of course, it's much more complicated than that."

Coyle was right.

A decade after cold fusion was first announced—and around the same time that hafnium was winding its way through government corridors—Peter Zimmerman, the U.S. State Department's scientific adviser and only PhD physicist, got up to give what would be a seminal speech at the annual meeting of the American Physical Society. Zimmerman, the scientist who would eventually introduce me to the hafnium bomb, took his stand against the scientific underworld in 1999.

Argumentative, opinionated, and never dull, Zimmerman's speech laid the groundwork for what he saw as an emerging battle: the war over pseudoscience in government. Zimmerman seemed to enjoy himself as he ran through the proposals that had crossed his desk over the past year, ranging from cold fusion and dubious claims of shrunken hydrogen atoms that could produce energy, to antigravity devices and perpetual motion machines, which violated the most basic laws of thermodynamics. The way Zimmerman saw the world, scientists were facing a barbarians-at-the-gate-scenario. Bizarre claims of unlimited energy and UFO technology were breaking down the doors of government, and skeptical scientists were being portrayed as intolerant elitists.

Be afraid—be *very* afraid, was his message.

"I am appalled to think that as the twentieth century draws to an end, two years from now, that we are seeing nonsense like that and we have very little defense and we as scientists don't do a damn thing very often to stop it," he said. "Well, I have decided that it stops as far as I can help it with me. . . . We're going to try to do a project in which we collect information about junk science funded one way or another for publicity purpose or anything else, by federal money.

"We are trying to go after them whatever way we can."

That speech was enough to make Zimmerman enemy number one of the scientific underworld. "Who is Peter Zimmerman?" asked the summer 1999 issue of *Cold Fusion Times*, a Xeroxed fanzine whose running banner declares its goal to "coldly go where no one has gone before." Another fringe science publication, *Infinite Energy*, accused Zimmerman of "terrorist-like activities." The cold fusionists, it turns out, were particularly angry with Zimmerman for his role in lobbying against plans to hold a conference on "free energy" at the State Department. Zimmerman successfully blocked the conference, which was going to feature a variety of fringe science, from State Department's Foggy Bottom headquarters. Government shouldn't be in the business of promoting pseudoscience, was his argument.

It may not be surprising that when Collins's devotees began showing up at Zimmerman's office in 1998 to tout the hafnium bomb, the physicist was highly skeptical. Zimmerman had long heard talk about isomers as a potent energy source for weapons, but had never taken it very seriously. With a PhD in nuclear physics, he was well versed in the scientific issues. His thirty-year career spanned the overlapping worlds of science and national security. But none of that had quite prepared him for the hafnium bomb.

It was shortly after the dental X-ray experiment that two men from the Pentagon's Defense Intelligence Agency—Fred Ambrose and Elliot Lehman—showed up at his office. No one knew for sure what Ambrose did, except that he seemed to revolve between CIA and DIA, and his main obsession was isomers. Lehman, similarly, had made a career following other countries' "directed energy" weapons, like lasers. This time, however, both men were insisting that hafnium weapons were right around the corner. Zimmerman listened politely, convinced it was all nonsense.

That wasn't the end of it, however.

The hafnium believers who had been coalescing at the Sandia lab in New Mexico also started lobbying Zimmerman to support the hafnium bomb. James Corey, the intelligence official from Sandia who had watched with suspicion the Russians giving lectures on the area, tried to convince Zimmerman of the threat posed by a fearsome isomer bomb. Next in line were Nancy Ries and Patrick McDaniel, who were not only promoting the hafnium bomb, but were actually handing out campaign buttons that read, "I believe in isomers." They paid a visit to Zimmerman's office in Washington, touting hafnium "as the best thing for weapons research since sliced bread," he recalled, adding that he wished, for posterity, that he had held onto one of their buttons. They also introduced another argument into the mix: because isomer triggering didn't involve nuclear fission, building a weapon using hafnium wouldn't violate the moratorium in place on nuclear testing.

That really got Zimmerman's attention.

The "I believe in isomers" campaign first hit Zimmerman just as he was preparing to take over his new job in Foggy Bottom as chief scientist of the Arms Control and Disarmament Agency, whose mission was both to promote arms control and to be on the lookout for new developments in weapons. As chief scientist, Zimmerman was responsible for preventing "technological surprise" in the weapons field: the idea that foreign countries could somehow develop technologies, like isomers, that would allow them to unexpectedly surpass the United States. Though the science of an isomer bomb seemed to him to be highly questionable and the promises vastly unrealistic, he couldn't stop thinking about the 1939 decision by the Naval Research Laboratory to ignore an Italian-born physicist, Enrico Fermi, who tried to convince the U.S. military that the fascists were working on a new weapon based on nuclear fission.

The military thought Fermi was talking science fiction; Zimmerman had a similar reaction to the isomer bomb.

"I had the science fiction reaction, and a rather bad science fiction at that," he recounted. "But what I wanted to know was that if I discouraged the Defense Department from funding it, I wouldn't be like the admiral who turned down Enrico Fermi in 1939."

6

DO YOU BELIEVE IN ISOMERS?

EVERY SUMMER SINCE 1959, a small group of the nation's top scientists quietly gather in sunny La Jolla, California to study issues critical to national security. Operating just out of public view, the scientists rank among the nation's elite and hold some of the government's highest security clearances. Through most of its history, the group, called simply JASON, has operated under contract to DARPA, although its annual studies span many areas of national security. Not unlike DARPA, the JASONs were born of Cold War–era concerns about science and national security. Also like DARPA, the group traditionally kept a low profile and has fiercely guarded its autonomy. Whether intentional or not, its actions have contributed to an aura of secrecy.

Even the name, for many years, was shrouded in mystery. A persistent rumor was that the group was named after the months of year they meet: July-August-September-October-November. As a group—they spell their name in capital letters, JASON, and the members are referred to collectively as the JASONs.

Where, then, does the name come from?

The reality, Marvin Goldberger, a founding member, said in later interviews, was that back in 1959 when the group was formed, a Pentagon computer had spit out the codename "Project Sunrise." Goldberger's wife didn't like the name, and suggested JASON after seeing a building that reminded her of a Greek temple; she thought it was an appropriate way to christen a group of men heading out on a new voyage, like Jason and the Argonauts of Greek mythology. The JASONs had their own noble mission: it was the dawn of the Cold War, and they were young scientists who wanted to contribute to national security. (They also, as some members later admitted, wanted to earn a bit of consulting money.)

Today, the JASONs are managed by a small office in northern Virginia at the MITRE Corporation, a federally funded research institute that provides high-level technical and intelligence services to the Pentagon. There is no public list of the JASONs and the JASONs' office in northern Virginia won't say exactly how many members the group has. They have no Web site. No public affairs. Their records, because they are a private group, are not accessible under the Freedom of Information Act, and a brief statement the program office provides to those few who inquire describes JASON as an "autonomous group" that is "governed by its own internal structure."

Often called upon to evaluate controversies beyond the scientific understanding of government officials, the JASONs have weighed in on items ranging from obscure technology to weighty policy issues, and their influence over the years has been subtle, yet arguably important. A 1966 report by the JASONs cast doubt on the use of strategic bombing to cut the Viet Cong's supply lines during the Vietnam War. The report, "Vietnam Bombing Evaluation by the Institute for Defense Analyses," is cited at one point in the Pentagon Papers, although the group is not referred to by name. The JASONs are simply called a "secret Defense Department seminar of 47 scientists," or "the cream of the scholarly community in technical affairs." Related JASONs work during the Vietnam War led to the creation of a sensor barrier that formed part of the quixotic McNamara line.

In more recent years, the JASONs have weighed in on scientific issues with highly politicized implications. In a 1995 report, for example, the JASONs concluded that low-yield nuclear testing wasn't necessary for the United States to maintain a robust stockpile of nuclear weapons, a recommendation that figured prominently in the Clinton administration's continued support for a moratorium on nuclear testing.

When hafnium hit Washington in 1998, the JASONs seemed a logical place to turn to resolve the debate. In fact, just a year earlier, in 1997, the JASONs, at DARPA's request, had looked briefly at a possible scheme for a hafnium bomb, as part of another report on new types of explosives. "There was a disturbing absence of a reality check in this proposal," the JASONs wrote in that report. "There was no scientific justification of how such a process might possibly be obtained with high enough efficiencies to be useful in any practical purpose."

Unless someone proposed some explanation for how such a bomb could work, any current proposal had "no merit at the present time."

But that report was before the 1998 experiment. With the hafnium campaign in full swing, Peter Zimmerman decided to hand the hafnium issue over to the JASONs. The group had the scientific clout and the objectivity to say, once and for all, if isomer triggering was bunk. The hafnium study was officially handed over to the JASONs for their 1999 summer meeting, and the isomer believers were invited to La Jolla to present their work.

Among the JASONs who worked on the hafnium study that summer were some of the nation's top physicists. Steven Koonin, a nuclear theorist and provost of the California Institute of Technology, chaired the panel. Other JASONs assigned to the hafnium review included a number of famous scientists, including Richard Garwin, who had helped design the hydrogen bomb, William Happer, a Princeton physics professor who had once headed the Department of Energy's science office, and Sidney Drell, a Stanford physicist and a longtime Pentagon weapons adviser.

Patrick McDaniel and Nancy Ries agreed to go to the JASONs meeting, but Carl Collins, for unclear reasons, dispatched James Carroll, his former student, in his place. That left McDaniel, Collins's longtime friend, to brief as a representative of the Air Force, while his colleague Ries presented for the Sandia Lab. But if McDaniel and Ries were briefing their "concept for an interesting weapon," as Koonin, the chairman, would later describe it, it was Carroll, standing in for Collins, who really took a beating at the hands of some of the nation's top physicists. Garwin, never known for modesty, grilled Carroll, repeatedly interrupting him until Koonin finally had to intervene and ask that the beleaguered scientist be allowed to finish.

As the various JASONs filed in and out of the room, Carroll presented the results of the dental X-ray experiment to a group of extremely skeptical scientists. Carroll would later describe the briefings in a diplomatic tone as "an eminent group of very accomplished people." But, he added, that doesn't mean a panel review of isomers could determine everything. Of the JASONs who attended, none really specialized in nuclear isomers, he said. Of the nine panelists, roughly a third had the technical expertise to really follow the nuclear physics involved, another third were at least interested in what he had to say, and the remaining third seemed unknowledgeable and uninterested.

But Carroll wasn't attempting to argue for a hafnium bomb. That was left to McDaniel and Ries, who got up to promote the isomer weapon, trying to take up at the point where the dental X-ray experiment left off. It wasn't enough to trigger hafnium to create a bomb; there had to be a way to spark a chain reaction similar to the way a match sparks a molecule of dynamite, igniting the next molecule, and so forth. In nuclear fission, neutrons spark the chain reaction, but in hafnium, there didn't seem any way to do it. Briefing the JASONs, McDaniel argued that another element could be put into a bomb to spark a chain reaction. The problem was, he didn't know what this element would be, and there didn't seem any proof that it would work.

Lev Rivlin, the Russian scientist who first patented the gamma-ray laser, had once proposed his own idea for "a nuclear chain reaction." The proposal was a bit farfetched, as even Rivlin admitted, because it required finding two different isomers that would somehow be a perfect match such that the natural decay of one would trigger the other. "Discovering of such a remarkable nuclides would be a very pleasant, but infrequent surprise," Rivlin once wrote. Another equally mystifying concept for a chain reaction was an "isomer combustion reaction" that involved hot, dense plasma or various methods of ignition using exploding wires or terawatt optical lasers, the Russian scientist suggested. Carl Collins had also once suggested his own "ignition concept," or an "electromagnetic analog of detonation." Collins's proposal seemed like even more of a stretch, presuming that a triggered isomer would give off secondary X-rays at precisely the right energy levels. None of those proposals for a chain reaction, including McDaniel's, had ever been written up in any meaningful formulation that could be scientifically reviewed and critiqued.

McDaniel, for his part, bristled at the mention of the JASONs, and accused Richard Garwin of orchestrating the attacks. Garwin, a big critic of the Strategic Defense Initiative, the Reagan-era missile defense shield, had an "axe to grind" and was taking it out on hafnium, according to McDaniel. "I don't want my mother's retirement being spent on some of the flaky schemes I've heard of," McDaniel said, but Garwin was taking a philosophical dislike to hafnium, and

Koonin, the head of the study group wasn't flexible on the science.

"Koonin didn't want to hear what we had to say," according to McDaniel. "He wanted to hear about the dental experiment. That JASONs report was right after the first dental experiment and we didn't understand it ourselves very well."

But even McDaniel acknowledged a major problem: Collins, lead investigator for the hafnium experiment, had refused to attend the JASON meeting. "Carl didn't want to have anything do with them," McDaniel said.

In 1998, Zimmerman had charged the JASONs to determine whether isomers could be made into a weapon over the next twenty years, looking at four principal questions: Did Collins demonstrate that an "enhanced decay rate," or triggering, really occurred? What was the physical mechanism that would allow the triggering to take place? Could enough hafnium be produced feasibly in the next twenty years to make it useful? Could a triggering mechanism be produced in the next twenty years?

The JASONs' conclusions, reached in July 1999, were damning on all four fronts. The JASONs determined that there was no clear evidence that Collins's group actually triggered the hafnium, there was no mechanism to initiate the sort of chain reaction required of explosives, and no feasible means to produce hafnium. The JASONs' study also attacked Collins's work, calling the experiment "poorly characterized and ill-described in the paper."

Where was the internal consistency? Where was the proof the "effect" was actually associated with hafnium-178 and not with trace elements mixed in with the source? The JASONs also challenged the statistical accuracy, wondering how such a small effect could lead Collins to such extraordinary claims?

There was another problem with the hafnium bomb, the

JASONs noted. The thirty-one-year half-life of hafnium-178 dwarfs that of the uranium or plutonium used in nuclear weapons, they pointed out. Thus, the thirty-gram hafnium weapon that McDaniel briefed to the JASONs would have 2000 curies of radioactivity. "Not only would this be difficult to shield against before explosion, but the radioactivity would be dispersed widely as a result of the explosion," Koonin wrote in his report. "In short, one would have constructed a radiological weapon; the same (highly undesirable) effect would be achieved far more easily by simply dispersing the isomeric Hf."

Yes, it was true that hafnium-178 has unusual properties, and as a high-density material, it could store a tremendous amount of energy, but as Koonin pointed out dryly in his letter to Zimmerman, "a jelly donut has about the same energy per gram as high explosive."

"Dear Peter," Koonin's letter to Zimmerman started, presenting the JASONs' review, including the briefings that took place that summer in La Jolla. After first summarizing the hard-core quantum mechanics, Koonin delivered the JASONs' verdict on the program. Collins results were "*a priori* implausible," and in a phrase that Koonin would repeat in later interviews about the review, he wrote, "extraordinary claims demand extraordinary proof." The Collins paper did not present such proof and neither did the briefings, according to Koonin.

Zimmerman had charged the JASONs with determining whether isomers could be made into a weapon over the next twenty years. The answer was no. The JASONs concluded that there was no clear evidence that Collins's group actually triggered the hafnium.

While dismissing the concept of an isomer weapon, the JASONs did point out that the possibility of triggered emission from hafnium, while unlikely, was an important issue for basic physics. They ended their report recommending that a definitive experiment be done at a proper X-ray source, such as a synchrotron.

Put in front a formal review, the isomer believers had faced their first defeat.

When I called up Steven Koonin, then the head of the JASONs, at his Caltech office in late 2003, I expected to be brushed off by a secretary, who politely took my message. But I had barely placed the phone receiver down when it rang with Koonin on the other end. Almost five years after the one-day JASONs briefing, Koonin, like a lot of people, was surprisingly eager to talk about hafnium.

The fundamental dilemma of the hafnium bomb, Koonin explained, was that even if it worked, it wouldn't be useful. All of the shielding that would have to go around the hafnium bomb to protect soldiers against the radioactivity would mean that it wouldn't be such a small bomb anymore. Koonin paused for a moment before adding: "It sure would make a great dirty bomb."

In either case, it wouldn't work, he argued.

"It's okay to be wild and crazy, but not everything you could imagine is worth pursuing," he said of the hafnium bomb. "There are other practicalities. It's not enough to say it's 'out of the box,' it's got to be plausibly out of the box."

Ironically, by the time Koonin sent his letter and final report to Peter Zimmerman, the Arms Control and Disarmament Agency had ceased to exist. At the behest of Jesse Helms, the Republican senator from South Carolina, the agency was rolled into the State Department. Shortly after that, the second Bush administration came into office, and Zimmerman, a Democratic political appointee, was moving over to a senior congressional staff position.

Zimmerman got the report as he was on his way out of the State Department, but with that one review, he thought he'd driven a stake through the heart of the isomer bomb.

Little did he know, the isomer bomb was just starting to build its momentum, beginning its roller-coaster ride from a controversial, but slightly obscure, 1998 experiment in a relatively unknown science center outside of Dallas to the inner sanctum of the military—the E-Ring of the Pentagon.

RICHARD GARWIN, DEPENDING on one's point of view, is either the most famous or infamous member of the JASONs. Tapped as a genius as a young student, Garwin at the age of twenty-three spent his summer at the University of Chicago designing a model of the hydrogen bomb for Edward Teller. He later went on to become a research fellow at IBM's Watson Center in New York, gained a reputation as an outspoken advisor to government on national security, and is widely credited as one of the founding fathers of modern satellite reconnaissance. But along with his contributions to the advancement of science is another talent: making people angry. Freeman Dyson, a fellow member of the JASONs, once described Garwin as one of the most unique physicists of the century, "a critic of unsound ideas, unsound experiments, unsound military projects."

Garwin's reputation, according to Dyson, was for "demolishing all three kinds."

It was in the summer of 2004 during the annual JASONs meeting in La Jolla, California when I finally managed to get Garwin, destroyer of unsound ideas, on the phone. We had exchanged a series of brief e-mails, which involved painfully complicated arrangements for how and when I could reach him, until finally, slightly exasperated, he simply wrote that he would call me.

Garwin, who had described the isomer bomb to me as "a technical and psychological problem," could not even understand why anyone was wasting their time on such a thing. Five years after that JASONs review, Garwin couldn't recall the briefers who spoke that day, and didn't seem to care much, because first, Garwin wanted to talk to me about Joseph Weber's gravity waves, an example of another unsound idea.

Weber, I learned, was a professor at the University of Maryland. Gravity waves are disturbances in the space-time continuum caused by colliding masses. Weber's fate collided with that of gravity waves when he set out in 1961 on an arguably quixotic quest to detect these interstellar disturbances using a bar of aluminum that would resonate when struck by the elusive gravity waves. That's where all the problems started, according to Garwin.

Gravity waves were first theorized to exist back in 1915 based on Albert Einstein's theory of relativity. It is one thing to theorize the existence of a phenomena and another thing entirely to detect it. Weber even had an ingenious setup, sort of, according to Garwin. Weber's experiment used a tiny piezoelectric crystal as sensors for the aluminum bar and the data was recorded on a chart similar to those used to show the seismic waves of earthquakes.

In 1969, Weber announced that he had detected gravity waves.

But there was a problem. Weber was practically the only scientist who could detect the waves. The effect that Weber reported was quite small, and other researchers, try as they might, weren't able to replicate those results. Weber's experiment was relatively small and simplistic, and other groups used bigger, more sensitive equipment. As time went on, Weber began to claim that others didn't understand or weren't capable of understanding his experiment. Garwin, who also attempted unsuccessfully to detect gravity waves, had a rather blunt take on Weber.

"His problem was that when he looked at the results of his experiment—and he was really no experimenter—he just believed the results."

Weber didn't stop with gravity waves, however. He moved on to neutrinos: small, elementary particles that can pass unhindered through matter. Like gravity waves, these phantom-like particles are extremely hard to detect. Unlike gravity waves, which were mostly of interest to egghead physicists, neutrinos were of interest to the Pentagon. If better ways to detect neutrinos could be found, it offered a host of tantalizing possibilities for the military, ranging from underwater communications systems to a new means to spot nuclear submarines. That led the military to show an outsized interest in Weber, when he claimed in a 1985 publication that he has used a sapphire crystal to detect neutrinos produced by tritium.

Many scientists were equally dubious of Weber's neutrino-detection claims.

Again, the question was not whether neutrinos could be detected, but whether they could be detected as easily as Weber had claimed. And this was where the esoteric musings of scientists ran into the paranoid fears of the military, because if neutrinos emitted from tritium were as easy to detect as Weber claimed, then the United States' fleet of nuclear-powered submarines could be located from hundreds of kilometers away. And in the mid-1980s, with the Cold War in full swing, the prospect of someone, like the Soviets, suddenly knowing with great precision the location of submarines carrying U.S. nuclear weapons was of serious concern. As a later JASONs report put it: "If it should happen that Weber was right, it would be a matter of life and death for the submarines."

Except Weber was not right. Like with his gravity-waves detector, he was very, very wrong. What Weber claimed were results twenty orders of magnitude larger—that is, over a quintillion times more—than what theoretical physics would predict possible. The JASONs, in fact, studied Weber's experimental scheme for a year. They concluded, with mathematical certainty, that the force exerted by neutrinos would be much too small to be detected by a crystal in a laboratory.

Did this put a stop to neutrinos?

No, some people simply dismissed the JASONs results as naysaying. "That didn't deter folks who are tantalized or hafniumized by the idea that they are in on the ground floor of something very new, that other people have ignored or gotten wrong," Garwin said. "That is a great motivating influence."

But the oldest argument in all those cases, be it neutrinos, or gravity waves, or hafnium, is the same argument that comes up in almost all cases of scientific controversy. Didn't scientists in the past deny the possibility of manned flight? Didn't they scoff at the idea of harnessing nuclear energy? What about all the other technological advances that were initially disputed? How do the scientists know what the future might or might not bring?

When I brought up that point, Garwin began a monologue that was, if there was no other way to put it, quintessential Garwin.

> All of this information that we use on which the world lives—semiconductors, ordinary cameras, anything you see—is the result of decades, sometimes centuries of an increase of understanding by science and incorporation of science into technology. And so, before fission was discovered in the 1930s— and artificial radiation in the 1920s—the understanding of how neutrons cause reactions became, not household knowledge, but among physicists, it was well known. That led to the ability to predict a chain reaction carried by neutrons. People had thought about that before without knowing much about it. Leo Szilard in 1932 invented the chain reaction. He started with, I think, $5,000 from the British government to look at chain reactions—finding an

element that would produce more neutrons than it absorbed. He started at the low end of the periodic table with hydrogen and beryllium. He ran out of money. Had he started at the other end, we might have had nuclear weapons earlier.

That finally brought Garwin back to the subject of hafnium. Collins's published results were based on experiments with questionable statistical accuracy: a 4 percent result with an acknowledged 2 percent error. Had Collins and his group challenged their data robustly? Had they performed random checks of the baseline, to see if the measurable effect was really from hafnium? It didn't seem like it.

Sure, Garwin said, if you look at hafnium, it stores a lot of energy. But in the late 1930s, scientists all over the country developed tools that allowed them to better understand nuclei and how they transition from one energy state to the next. "I think it's marvelous to see this rotational band; to be able to identify which transitions are involved and to be able to calculate it," he said.

Triggering that transition using low-energy photons is simply not possible, he said. Enough is known about the nucleus of hafnium to make that determination. So, what did Garwin think of the idea of mass-producing hafnium?

"You don't need the hafnium; you don't need to spend hundreds of millions of dollars on hafnium production facilities to find the missing link in this process," he said.

JOHN SCHIFFER, A senior physicist and associate director of Argonne National Laboratory's Physics Division definitely did not believe in isomers. In 1989, Schiffer had been one of the twelve scientists assigned to a government cold-fusion review panel, and had seen firsthand what could happen when the scientific process goes berserk. Reading Collins's dental X-ray article when it came out in 1999, all the Argonne physicist could do was shake his head at the incredible claims. "When the results of that first paper came out, it seemed strange, and many nuclear physicists said it just couldn't be right," Schiffer later recalled to me.

Schiffer had an old world accent, and his manor and tenor reminded me of the Central European professors back in college who were constantly mortified by the ignorance of their students. Among colleagues, Schiffer had a reputation as a world-class scientist, albeit a tad blunt, with a habit of calling people fools to their face. But Schiffer was not the only scientist to take note of the Collins paper; the results had caused a stir among the community of scientists familiar with isomers, and particularly among those acquainted with Collins from the days of Star Wars.

"There were some comments published, criticizing the paper, but most people just talked at it as something not to be taken seriously," Schiffer said.

But a year and a half after the initial results were published, John Becker, a senior scientist at the Livermore lab, began to reach out to some of his fellow nuclear physicists. Becker's explanation for what would later become called "the Argonne experiment," was simple. "We pay attention to things that have national importance," he said.

That reasoning, in a sense, was not surprising, since a large part of the nuclear weapons laboratories' mission is to prevent technological surprise, and Collins's claims were certainly surprising. The lab, according to Becker, even began to get calls from defense companies wanting to know if there was any validity to Collins's experiment.

The initial Collins paper surprised everybody, Becker said, because the results were not at all in line with what was known about the physics of isomers. "The thing is, people know how to do this," Becker said. "You can use photons to excite nuclear states and that also means you can de-excite them, but the probability is very small."

The question was over the amount of energy needed to force the decay, he explained. The probability that Collins could do what he claimed to do using a dental X-ray machine was infinitesimally small.

"The issue with Collins's experiment—the rates he claims—he called it cross sections—are so high that they're well beyond any physics calculations you could make," Becker said. "It demanded new physics."

Determined to do an experiment that would either confirm or contradict those results, Becker eventually put together a group of scientists that included Schiffer and a dozen other researchers from three of the nation's leading government labs: Argonne National Laboratory in Illinois, Lawrence Livermore, and Los Alamos. Collins's 1998 experiment included a hodgepodge of Eastern Europeans, graduate students, and military scientists, while the Argonne collaboration included top names in nuclear physics from the some of the most prestigious labs in the country. Using funding they obtained from the Department of Energy, they prepared an experiment designed to replicate Collins's results.

It was, however, an odd experiment to conduct, Jerry Wilhelmy, a Los Alamos physicist would later recall. "There was really very little in it for us," he said. "There were only two outcomes. Either Collins could be correct and this would be very exciting since it meant there was a lot of new physics which wasn't understood, but, if so, the credit would and should go to him. But if Collins were wrong, which certainly most of the group thought he was, colleagues in the scientific community would question why we wasted our time pursuing hafnium when it was quite obvious from all we know that it couldn't work."

In the end, they considered the experiment something of a public service.

One memorable highlight, according to Wilhelmy, was when Donald Gemmell, another scientist involved in the work, showed up at the experimental hall at 3:00 A.M. with his bagpipes to serenade the other scientists. "This was certainly a first in my scientific career," Wilhelmy said. That moment of levity somehow seemed appropriate for what some considered a fool's errand: repeating an experiment that most scientists thought could only lead to nothing.

Nonetheless, using hafnium produced at Los Alamos and the powerful X-ray source at Argonne, the team of scientists attempted to reproduce Collins's results. Collins, of course, had used a dental X-ray machine. The Argonne team used an advanced X-ray source the size of a football field, which was one hundred thousand times more intense than Collins's dental X-ray. According to the scientists who participated in the Argonne experiment, if Collins's results were correct, then their team should have seen a much bigger signal. But when they turned on the X-ray, they saw nothing.

In a sense that wasn't really surprising, since no one seemed to think that Collins's results were correct, anyhow. The laws of physics were still comfortably in place.

In May 2001, Becker, the head of the Argonne experiment, spoke briefly at an isomer conference held in Telluride, Colorado to present those results. It was a conference organized by Collins's colleagues and devoted primarily to Collins's and his colleagues' work, so Becker was a bit of an outsider, particularly since his presentation was a refutation of Collins's experiment. Collins and Becker even spoke in person, and both men described it as a civil conversation.

At that point, the Argonne group had already performed their first experiment to the specifications Collins provided in his initial article in *Physical Review Letters*. Collins, however, took the opportunity to tell Becker that the Argonne experiment was meaningless. The Texas group's latest work was confirming that even lower-energy photons, around ten thousand electron volts, or 10 KeV, could trigger hafnium. Argonne hadn't tested at those low levels, he pointed out.

This was convenient for Collins, who was now able to claim that the Argonne experiment, though conducted the way Collins had described in his first paper, hadn't been quite right. Collins, in the meantime, had published another, longer paper, in 2000, this time in *Physical Review C,* describing a new hafnium triggering experiment, though still using the dental X-ray machine.

Not described in the article was the unique way Collins's group went about proving this new discovery. For this experiment, they used sheets of Reynolds aluminum foil to filter out higher-energy photons. Compared to the dental X-ray, that was almost high-tech. In either case, this new information wasn't in the first article, according to Collins, because his group only learned that after repeating the experiment, and filtering out the higher-energy photons.

But one of the problems was that Collins didn't tend to give a whole lot of details about his experiment. In speaking about that early exchange, John Becker, the Livermore scientist, seemed at a loss for words. "How do I want to say this?" Becker said with a sigh. "I think his papers are very difficult to read."

Nonetheless, the Argonne group went back and did a second experiment a year later to look at the low-energy levels Collins described. Again, they found nothing. Again, Collins found issues with their experiment. This time, he claimed new problems: their target didn't contain enough hafnium, their detectors weren't looking for the right energy levels, and the X-ray flux was all wrong. What followed were a series of publications and counter-publications in American and European journals. In 2001, Donald Gemmell, a physicist from Argonne, wrote one journal warning the editors that they were in danger of promoting a new "cold fusion."

In almost didn't matter. By that point, Collins wasn't being published anywhere except in Russian journals. And Becker's statement—that Collins's results required "new physics"—eventually became the mantra behind which many nuclear physicists would rally. It wasn't just whether one experiment or another showed the results, but the very physics that Collins claimed were improbable. Becker's group had produced what they considered to be a "textbook" experiment on isomer triggering, and the critics thought that with Argonne's results in print, the 1998 experiment was destined for the scientific dustbin. That might have been true, if it weren't for the military.

Collins's work wasn't just picked up and promoted by the Air Force, it also was finding proponents in other parts of the national security establishment. Despite a negative report, two counter-experiments, scientific journal articles doubting the theoretical and experimental basis of his work, isomers were racing full speed ahead. The criticism was proving no more than a minor speed bump. Isomers were about to hit the Capital Beltway.

7

HAFNIUM COMES TO WASHINGTON

TEMPERATURES THAT EARLY August morning in Washington
were already creeping above 100 degrees. It was the summer
of 2005 and I was riding in a minivan with Lloyd Feldman,
the assistant director of an obscure Pentagon division called
the Office of Force Transformation. Feldman's office was
one of several in the Pentagon designed to ferret out new
and unconventional approaches to warfare—or in Penta-
gon-speak, "transformational" technology. We were on our
way to visit one of the division's latest inventions: a vehicle
equipped with a device that could shoot adversaries with an
invisible heat beam. The Pentagon was preparing to send it
to Iraq. Our conversation slipped from bouncing lasers off
satellites, to DARPA, and then to nonlethal weapons, until
we hit a bump in the road and suddenly we veered off onto
the subject of the merry band of psychics who once worked
for the Central Intelligence Agency.

Dubbed Stargate, the U.S. government's forays into
"remote viewing," colloquially known as extra-sensory perception, or ESP, are often considered one of the most embarrassing episodes in official fringe science. Beginning in the
early 1970s, the intelligence community began hiring psychics
to describe objects in remote locations. What better way to
win the Cold War than to pinpoint with precision the location of Soviet missile silos, nuclear submarines, or sensitive
military installations?

The central figure in the U.S. psychics program was an unusual scientist named Hal Puthoff, who convinced intelligence officials that psychic phenomena could be proved. Recounted with profound credulity in Nick Cook's best-selling book, *The Hunt For Zero Point,* the CIA's obsession with psychics—not unlike the hafnium bomb—emerged from paranoid fears of Russian superiority. In other words, the CIA wanted to close the psychic gap. According to Cook's account, Puthoff, then a scientist at the California-based Stanford Research Institute, came upon a description of Soviet investigations into ESP among ordinary fauna. Plants, the Soviets theorized, could somehow communicate with each other using something called *tachyons,* hypothetical particles that travel faster than the speed of light. Apparently, scientists in the United States were so behind the Cold War curve that they weren't even aware that plants could think, let alone that they could read the thoughts of the flower in the next pot over.

World domination might be only a step away.

It wasn't quite the Manhattan Project, but Puthoff set out to study the bourgeoning field of parapsychology and his proposal worked its way up the bureaucratic chain and into the hands of the U.S. intelligence community, which for the next fifteen years funded psychics as a secret program under the vaguely scientific title of "remote viewing." For years, psychics virtually roamed the Soviet Union in search of military installations, without ever leaving the confines of sunny California. Or at least that's what they tried to do. Fortunately for national security, the government was also building multibillion-dollar satellites and sensors using state-of-the-art technology. As one government scientist quipped: "Isn't it a wonderful concept? The two arms of the U.S. submarine detection effort: A $20 billion line of sound detectors scattered all over the oceans and a nut sitting in a darkened room in Palo Alto doodling maps."

Were it not for the massive number of documents eventually released to the public following the disclosure of Stargate, no one likely would have ever believed that high-level Pentagon and military officials were shuffling memos back and fourth as late as 1994, seriously debating the merits of ESP and attempting to get more money to keep those efforts alive. By the time the CIA's psychics were "outed" in 1995, no one really was that interested in spotting Soviet installations, and in any case, satellites had proven much more reliable. The program was unceremoniously terminated.

While I knew the idea of psychic phenomena still held some appeal in Pentagon circles, I was caught off guard when Feldman, who struck me as one of the brighter defense officials I had shared a ride with, expressed interest in ESP. Musing about the military's fascination with psychics, he suggested there might be something useful in it.

"You're kidding?" I said, eyeing the offices of the various defense companies as they breezed past us. They were tall, featureless buildings with reflecting windows—anonymous except for the insignia attached to the front. I tried to imagine that in the depths of one of those cubicle-filled boxes, some Beltway bandit was busy writing up a new proposal to study psychic phenomena. That clever employee would figure out that if he shopped the proposal around to enough offices of the Pentagon, he would eventually find a government official willing to give it some money. Maybe that official would be Feldman.

Or maybe not.

"I'm not suggesting that I believe in psychic powers," Feldman quickly added. "But maybe there is some weak scientific phenomenon at work that could be measured and studied."

I thought of Hal Puthoff with his plants and tachyon particles. In fact, much of what Puthoff did while working at the Stanford Research Institute was precisely what Feldman suggested. He had studied possible paranormal phenomena in a manner that maintained at least the veneer of scientific credibility. He had attempted to measure and quantify ESP. Of course, there was a problem with those early studies—a problem that persisted until 1995 when the program was shut down amid a barrage of embarrassing publicity. How do you measure the existence or nonexistence of a psychic phenomenon? The psychics were drawing pictures of objects they "viewed," but at the end of the day it was up to a human being to decide whether they had successfully described the object. Was it a barn, or a missile silo? A submarine or a whale? Independent scientists brought in to evaluate the program in the 1990s couldn't even agree among themselves on how to evaluate the data.

There was another vexing problem. Most scientists agreed there was a certain element of chance involved in all the tests. If asked to view an object remotely, a would-be psychic might just get lucky. But how many correct answers are just lucky guesses? How much of a statistic beyond those lucky guesses is meaningful proof of psychic abilities? At one point, the scientists reviewing the data decided to set the baseline for "lucky guesses" at 20 percent. To prove that the psychics indeed had actual paranormal abilities, they had to get better than 20 percent correct. That's where things got very complicated.

The problem is known as the *signal-to-noise ratio*, Feldman explained to me. In science, this term is often used to describe the ratio of "true," or real data, to that which is false or irrelevant. Noise, or false signals, could be almost anything—for an X-ray experiment it might be a problem in the detectors, and with psychics it's the lucky guesses. The goal of any scientific study is to increase the signal relative to the background, which helps dispel any doubt that the result could be attributed to chance or error. To prove a psychic phenomenon, Feldman asserted, scientists need to boost that signal.

Listening to Feldman, I thought back to the original dental X-ray experiment; Collins claimed just a 4 percent triggering of the isomer, barely above the statistical error of 2 percent. When I asked Feldman about isomers, he said that he had heard of them, but it really wasn't his area of expertise. If isomer triggering were real, the scientists should also try to boost the signal, he said.

Like with the psychic phenomenon, you need to know if you're seeing a signal. Did Collins see a signal, or was it just noise? Did he, as physicist Richard Garwin suggested to me, simply set his baseline at the wrong place?

Feldman paused to look out the window at the passing landscape, and smiled. The van was racing beyond the outskirts of the Washington Beltway. The massive buildings of the Pentagon's contractors—Lockheed Martin, SAIC, Northrop Grumman—were receding into the distance.

"You know, life is background noise," he said. "Making sense of life is all about boosting the signal."

EVEN AS COLLINS'S results were being greeted with skepticism in the scientific community, isomer enthusiasts had begun making the rounds of Washington's military and intelligence agencies seeking out support. And chief among them was Forrest "Jack" Agee, an Air Force scientist and one of the principle believers in hafnium. Agee believed in isomers and he didn't seem too concerned about what the JASONs or anyone else thought.

By 1999, as the head of the physics division of the Air Force Office of Scientific Research, Agee also was finally in a prime position to help Collins. The dental X-ray experiment gave him the perfect opportunity. On the basis of that alleged breakthrough, Agee began funding Collins, providing about $400,000 a year for experiments undertaken by Carl Collins, James Carroll at Youngstown University, Patrick McDaniel at the Air Force, and Hill Roberts at SRS Technologies in

Alabama. The hafnium believers finally had a sponsor in the capital.

And what was particularly odd—and, to Department of Energy scientists, simply unacceptable—was that Agee, who controlled a hefty $55 million in physics funding, became both a sponsor of isomer work, and simultaneously a beneficiary; he was being named as a coauthor on the scientific papers being produced by Collins and Carroll. For Collins's critics, it was a break with convention, but it apparently didn't raise any eyebrows in the Pentagon.

Not only was Agee supporting Collins's work in the United States, but the London-based division of the Air Force Office of Scientific Research was supporting Collins's Eastern European colleagues, including the Foundation for Induced Gamma Emissions, the research institute run by the Popescus, Collins's old friends from Romania. Other grants were sprinkled around to Collins's colleagues in Ukraine and Russia.

But there was one small problem for the hafnium scientists. Agee was supporting both Collins and Carroll, who were no longer working together, and Collins wouldn't let anyone else use his (or actually his wife's) hafnium sample. Without another source of hafnium, no one else could repeat Collins's experiment. Agee himself in 1999 solved that problem, however. Agee had the Air Force provide a contract—a small-business award of $150,000—to buy another hafnium sample from Los Alamos. The contract was provided directly to Hill Roberts's company, SRS Technologies.

Over the next few years, isomer believers began to coalesce around Collins's work. Supporters—from the CIA to Sandia—touted his results as confirmation of what they always suspected to be true: that nuclear isomers would make for a fantastic weapon. So while Collins's work between 1999 and 2001 was becoming increasingly marginalized in the scientific community, the believers were trumpeting his results in Washington with considerable success.

By 2000, the hafnium supporters were gaining momentum, driven by a new zeal, if perhaps somewhat less-than-compelling data. Collins and his former student Carroll weren't getting along, but they were both being funded by the Air Force. Fred Ambrose, by then at the CIA, was pushing hafnium in Washington, while Patrick McDaniel was working on hafnium in the Air Force Research Laboratory's directed-energy division, and Nancy Ries, at Sandia, was also involved in funding isomer research. But the isomer believers had found a new friend: Martin Stickley, a government scientist working in London for the European branch office of the Air Force's Office of Scientific Research. Stickley was funding Collins's Eastern European (mainly Romanian) colleagues, and he, too, had become a believer.

In May of 2000, Stickley, the latest to enter the small world of isomer believers, invited a select group to meet at the Air Force's European Office of Aerospace Research And Development's Edison House in London. Twenty people attended the meeting that was nominally dedicated to developments in the gamma-ray laser. But what quickly emerged from the meeting was a far more ambitious agenda. There was a sense that isomers were set to takeoff with substantial government funding, and the scientists involved wanted to position themselves for a possible boon.

Carl Collins, for starters, proposed that his Center for Quantum Electronics could be established as the National Center for Isomer Research, a sort of leading clearinghouse for all isomer work. Lev Rivlin, the Russian scientist who had first patented the concept for a graser, also had his own proposal for a full-scale experiment that would cost $3.65 million.

But to hit prime time, isomers needed a home, and despite the battering isomers took in the IDA report, and earlier with the JASONs, the best place for hafnium was still in the confines of the Beltway. Rejected by the scientific journals, the mainstream nuclear physics community, and even by weapons designers in the Pentagon, isomers needed a place that was resistant to oversight. They also needed money. While Agee continued to fund Collins and his colleagues, the amounts the Air Force could provide were still relatively small. To really hit prime time, isomers needed some place with big money and protection from outside scrutiny. That place was increasingly likely to be Donald Rumfeld's Pentagon, and more precisely, DARPA.

ON MARCH 4, 2002, in the middle of the battle of Tora Bora in Afghanistan, Secretary of Defense Donald Rumsfeld announced at a Pentagon press conference that the U.S. military fighting in Afghanistan had used something called a "thermobaric weapon."

The name alone sounded ominous.

Thermobaric, an amalgam of the Greek words for heat and pressure, is a slow- burning incendiary warhead designed to create lethal overpressure by sucking up oxygen from an enclosed space—like a tunnel—to create a powerful vacuum. The overpressure effects of a thermobaric weapon can cause a gruesome array of injuries. Internal hemorrhaging, collapsed lungs, eyes torn from eye sockets, and ruptured eardrums were all reported as injuries linked to the thermobaric weapon. Much to the outrage of human rights groups, the Russian military also employed a type of thermobaric weapon during its own war in Afghanistan, and again a decade later, during the first Chechen war, according to media reports.

Thermobaric weapons weren't the only things the Pentagon was interested in, however. On the heels of the ongoing war on terror, the Bush administration in 2002 unveiled a few details of its classified Nuclear Posture Review, a congressionally mandated assessment of the U.S. military's nuclear stockpile. The report outlined the administration's vision of a smaller, yet more robust nuclear force. And much to the frustration of arms control analysts, it proposed developing new nuclear weapons that could be used to penetrate deep into hardened and buried bunkers. Finally, it recommended that the military identify new and more powerful conventional explosives that could achieve effects similar to nuclear weapons.

In the early days of the war on terrorism, the Pentagon began touting a number of powerful bombs, some new, while others were older weapons adapted for new uses. The awesome thermobaric weapon was followed by the Daisy Cutter, which was really just a Vietnam War–era device used to clear helicopter landing zones of dense jungle overgrowth—the fifteen-thousand-pound bomb created a mushroom-cloud effect, which the Pentagon hoped would have a powerful psychological effect on unsophisticated enemy forces that might confuse it with a nuclear weapon. And then there was the Massive Ordnance Air Blast, or MOAB, which military officials gleefully dubbed the Mother of All Bombs. The twenty-one-thousand-pound bomb was the largest conventional weapon in the U.S. military's arsenal. The Pentagon released pictures and video clips depicting tests of these new weapons, demonstrating the world's most advanced firepower. Officials told the media that the primary purpose of MOAB was in fact "psychological."

Donald Rumsfeld liked big bombs. He did not, however, like scientific advisers, at least not those who disagreed with him. The JASONs—the group of outside advisors who had panned the idea of an isomer weapon back in 1999—were unceremoniously dropped from DARPA's payroll in 2002 after a bitter dispute over their membership. According to several of the JASONs, the conflict started when DARPA requested that the group accept into its ranks three administration-selected favorites, including a Silicon Valley CEO.

The order, according to one version of events, came straight from Secretary Rumsfeld.

In effect, Rumsfeld wanted to make the JASONs a bit more like the Pentagon's other advisory panels. The Defense Policy Board and Defense Science Board are similar in purpose to the JASONs—they provide the Pentagon leadership with independent advice and counsel. But in more recent years, those panels have become dominated by what scientists call "industrialists," and what in the Beltway are simply called "defense contractors." In other words, consultants with strong ties to defense companies.

In an interview with *Fortune* magazine, shortly after the JASONs dispute, Rumsfeld defended his attempt at packing the scientific court.

"I tried to put a couple of people on a science activity in the department some time back—a couple of young Silicon Valley 30-year-old types," Rumsfeld told the magazine. "And it was considered sacrilegious to bring those folks into . . ."

Here Rumsfeld paused to lower his voice dramatically, "'. . . this distinguished body of people who have been on this particular activity for many, many decades."

The JASONs, who had always chosen their members by internal agreement, balked at Rumsfeld's attempt to bring in loyalists and the contract was dropped. After over forty years of summer meetings in La Jolla, the group was about to be disbanded. While the JASONs had normally operated in relative quiet, a furor arose when DARPA dropped their contract, leading many scientists to decry publicly the loss of what some considered a national treasure. Realizing the potential for embarrassment, a compromise was struck and the JASON contract was moved from DARPA to the Pentagon's Office of Defense Development Research and Engineering, a once-powerful division of the Defense Department whose role had long since been diminished. The new home was a good one, according to the JASONs, but the group's role in DARPA projects was diminished.

Rumsfeld offered his own assessment of the outcome: "We didn't win in that instance, but you just have to be persistent in this town."

As if taking a play from Rumsfeld's book, the hafnium supporters were indeed proving to be persistent, because by the time Martin Stickley showed up at DARPA with his hafnium bomb proposal in hand, the dental X-ray experiment had been refuted by two major government studies and two counter-experiments. But Stickley didn't seem to care. It was also convenient for him, perhaps, that the JASONs, who had panned the hafnium weapons just a few years prior, were no longer associated with DARPA. It would have been odd, even for an independent agency like DARPA, to fund an ambitious military program that its own independent scientific advisors had dismissed.

The JASONs being booted out of DARPA had nothing directly to do with the hafnium study, but the timing was certainly fortuitous in 2002, as Stickley was preparing to return to the agency. According to Hill Roberts and Patrick McDaniel, Stickley's primary interest in moving back to DARPA was getting the isomer program rolling.

In the middle of 2002, DARPA quietly moved funds between accounts. Hafnium became an isomer-weapon project, with no peer review, no oversight, and no questions asked. Just a bit of money mysteriously transferred between accounts with an addendum explaining the change. While there were initial reports that DARPA in 2002 had started two isomer projects—one to look at a bomb and another at a battery—within a few months, DARPA ditched the battery idea and the isomer weapon quickly took precedence. The hafnium project, according to a DARPA statement to Congress, was started as an energy program, but "was reoriented once its potential explosive power was discovered."

In its official statement on the goals of the isomer program, the agency made clear it had just one mission for isomers. "DARPA is interested in SIER [Stimulated Isomer Energy Release] because, if the effects are real, and we can demonstrate that the isomer can be produced in practical quantities, it may be possible for DoD to develop a bomb that weighs 50 pounds but that has the explosive force of a 2,000-pound bomb."

Around that same time, the Pentagon quietly added nuclear isomers to its Militarily Critical Technologies List, an official compilation of materials deemed essential to maintaining U.S. military superiority. No one knew who added isomers, although one report tied it to a private contractor. According to the new entry, nuclear isomers, and hafnium-178 in particular, "has the potential to revolutionize all aspects of warfare."

While the report warned that technical barriers might exist, it continued with the following exhortation: "We should remember less than 6 years intervened between the first scientific publication characterizing the phenomenon of nuclear fission and the first use of a nuclear weapon in 1945."

8

SCARY THINGS COME
IN SMALL PACKAGES

THE ICONIC IMAGE of the mushroom cloud, the telltale sign of a nuclear explosion, has burnished itself into modern culture as a symbol of military might as much as an object of popular kitsch. It's adorned political ads, it inspired Salvador Dali, and in the 1950s, it was used to sell everything from shoes to cereal. It can also frequently be found in certain government offices. I once asked a former Pentagon official why so many nuclear weapons scientists hang pictures of the mushroom cloud on their wall. The answer I got was as beautiful in its prose and as it was disturbing in its substance.

"You must understand that at one time nuclear was the glamour stock and a source of considerable pride for those of us who were card-carrying members," he wrote me. "While the glamour is gone, it is still a source of pride for me and most of my erstwhile colleagues. Owners of identical sports cars and motorcycles often flash their lights or wave to each other—I suppose to signify that they are of like breed. We telegraph our nuclear 'breed' by the explosions on our wall."

In that scientist's office, for example, was a picture of Ivy Mike, the first detonation of a thermonuclear bomb. That 1952 test touched off a 10.3 megaton blast at Enewetak Atoll in the Pacific Proving Grounds, creating a fireball over three miles wide and a mushroom cloud that rose to over a hundred thousand feet. The island of Elugelab, where the bomb was set off, simply vanished into the ocean. Yet the picture of Ivy Mike, with it its soothing colors and geometric symmetries, was somehow strangely serene, he said.

"Could an artist have captured any better the ingenuity of man in rivaling the power of the ancient Gods and their unforgiving vengeance for anyone who would anger them?" the scientist mused.

For almost as long as isomers have been around, scientists have imagined ways to use them in powerful weapons. There is even a standard Pentagon briefing slide that demonstrates the comparative explosive power of various weapons—both theoretical and real—including one based on isomers. The chart begins on the far left-hand side with good old-fashioned chemical explosives, like TNT, and ends on the far right with futuristic-sounding "antimatter" weapons. Antimatter, so the theory goes, could eventually be the most powerful weapon known to man if anyone could figure out how to produce more than a few atoms of it.

In between those extremes are nuclear isomers, and then nuclear fission and fusion.

To better fathom the fascination with isomers—and the isomer bomb—it also helps to understand nuclear physicists' fascination with nuclear weapons. The ability to harness the elemental forces of nature confers a special status. Like the elite cadre of scientists who worked on nuclear weapons design, the isomer enthusiasts belonged to an exclusive club, albeit one of a slightly more offbeat nature; a growing scientific underworld was rapidly developing around isomers.

Fred Ambrose, the mysterious intelligence official obsessed with isomers, was also back in action. In 2001 Ambrose appeared at the Livermore lab and wanted to discuss the hafnium isomer bomb, which he was concerned was being developed in Russia. Weiss was amused. Weiss told Ambrose that the physics didn't add up. "In principle, you could make a bomb. In practice, it doesn't work," he told him.

But Ambrose was undeterred, and asked Weiss to invite James Carroll, Collins's former student, to Livermore. Ambrose explained that he was helping Carroll by funding one of his post-doctoral students, and wanted them to work more closely with Livermore. Weiss agreed to help make contact, but Carroll politely declined two separate invitations, both times saying he was too busy. Weiss was relieved—Livermore had done as Ambrose had asked them, and there didn't seem anything more to do.

Recalling Fred Ambrose's enthusiasm for hafnium bomb, Weiss chuckled. "Fred is a true believer." But Weiss was quick to point out that Ambrose's motivation was out of patriotism. Weiss held most of the hafnium believers in disdain, but Ambrose really did believe he had the best interests of the country in mind.

Ambrose was busy trying to find support around Washington for his pet cause, and he wasn't the only one. By the time he cornered George Ullrich in the CIA cafeteria to try to get him to invest in isomers, Ambrose had become just another member of the hafnium club.

GEORGE ULLRICH WASN'T sure if he believed in isomers, but back in the late 1990s, as one of the Pentagon's top weapons scientists, he was definitely excited about the long-lived metastable nuclei. Ullrich was intrigued by energy and energy storage of all types. That's what weapons are about. If you can store more energy, you have potentially more energy to release. And whether it was ball lightning—the unexplained phenomenon associated with thunderstorms—or antimatter, Ullrich was interested in energy.

Several years later—and after Ullrich had left the Pentagon—I sat in his Virginia-based office as he recounted, with amusement, how hafnium landed on his desk. He offered me a soda, sat back in his chair, smiled, and began to talk.

Ullrich's interest in isomers began with what he described as something like a cross between fishing expeditions and shopping trips to Russia back in the mid-1990s when he was a top official in the Pentagon's Defense Nuclear Agency. The Berlin Wall came down in 1989, and by 1991, the Soviet Union had collapsed in an economic and political morass. The secret world of Soviet nuclear science suddenly burst wide open, and impoverished physicists—mostly from Russia, Ukraine, and Belarus—were desperate for funds and eager to sell the only commodity they had: knowledge. Various parts of the U.S. government began giving money to the scientists; part of the goal, of course, was to keep Soviet nuclear scientists employed, and thus, less tempted to take jobs in countries interested in acquiring knowledge about nuclear weapons. Another motivation was far less altruistic. The United States wanted to find out what the former Soviet scientists knew.

Beginning in 1993, Ullrich and his staff began traveling to Russia with the offer to fund former Soviet scientists to conduct open research paid for by the U.S. government. What the Pentagon found, according to Ullrich, was that when it came to nuclear weapons effects, Russians knew pretty much what the United States knew. But in reviewing the open scientific literature, Ullrich and others realized the Soviet *were* ahead in one important area: advanced energetic materials. In other words, conventional materials that offered potentially more punch than your typical chemical explosives. "What they were looking for were things that had energy densities up to 100 times that of TNT. That was the gold standard," Ullrich said.

It was still desperate days for post-Soviet scientists, who, once showered with attention, suddenly found themselves working at institutes unable to pay salaries, or even the electric bill. Ullrich described walking into a conference facility, expecting a dozen or so hopefuls to shows up, only to find over a hundred eager scientists ready with proposals in hand that ranged from the offbeat to the fascinating. "We got everything from cold fusion to some very interesting chemical compounds and other forms of exotic energy storage," he said.

One area that immediately intrigued Ullrich was isomers. The Soviets, it turned out, had done a fair amount of work on isomers; much more than their American counterparts, and Ullrich's office ended up funding a few Russian scientists to work on them. "It was very inexpensive," Ullrich explained—only $10,000 or $20,000 per grant.

Ullrich, like many scientists, had also been intrigued by isomers in the early 1970s when he was doing his PhD in physics. The idea of using nuclear isomers has been around for so long because the energy storage is quite large, Ullrich told me. "You're within two orders of magnitude to nuclear fission," he pointed out.

Also sitting in Ullrich's office that day was Michael Frankel, his former deputy at the Defense Nuclear Agency. Ullrich began talking about all the various nuclear phenonena that were not completely understood, and Frankel, who was wearing a tie with the periodic chart of elements, suddenly jumped up to the chalkboard to illustrate some of the things Ullrich was talking about. There were isomers, of course, but there was also ball lightning, the sporadic reports of a luminescent discharge sometimes seen during lightning storms. There was antimatter, the material that would explode when it came in contact with matter. And then, of course, there was cold fusion.

"Cold fusion?" I asked.

Ullrich nodded his head and went over to his bookshelf, pulling out a weighty tome called *Excess Heat: Why Cold Fusion Research Prevailed*. He placed it on the table in front of me, and I picked it up, flipping through the pages. It was dated 2002.

There were reports that cold fusion experiments—once dismissed as pseudoscience—were reproducing more reliably, Ullrich explained. That didn't mean for certain cold fusion was a nuclear reaction, but simply that there was something there that might be worth examining.

"You can't be overly cocky, however 'lunatic fringe' the ideas might seem." There's nothing flaky about funding ideas that are indeed high risk, he added, but you also have to be skeptical about outrageous claims.

When Collins and his supporters showed up to brief Ullrich's deputy, Frankel, he listened to what they had to say. While highly skeptical, Frankel decided it would be worth it to give Collins $60,000—"chump change" by Pentagon standards he told me—but enough, he hoped, to get Collins to use something better than a dental X-ray device. It didn't work. Collins continued to use dental X-rays, and Frankel presumed the money was spent on salaries for postdoctoral students.

But in 1999, with news of the dental X-ray experiment spreading rapidly through the Pentagon, the Defense Nuclear Agency, newly re-minted as the Defense Threat Reduction Agency, was planning to hold a conference dedicated to isomers as the hot issue in weapons research. And while Frankel and Ullrich both had their doubts about Collins, they were still quite enamored with isomers. There was a feeling that something big was on the horizon. And then came a call from Livermore. It was one thing to be warned that Pentagon officials were getting excited over nothing, but the Livermore message got their attention. The scientists there were basically warning that the Pentagon was about to be taken for a ride.

Ullrich got what he would describe, quite simply, as "cold feet." Things just didn't seem quite right, he decided, and cancelled the conference. That didn't stop the intelligence community, because in 2001, Elliot Lehman, an official at the Defense Intelligence Agency involved in directed energy weapons, and Fred Ambrose, who was bouncing between the DIA and CIA, held their own conference on isomers. Lehman and Ambrose were still trying to drum up support in Washington.

Frankel, who attended the conference, recalled listening to James Carroll, Collins's former student. There were some things about Carroll's talk that struck Frankel as a bit odd. For starters, Carroll alluded to the French hafnium triggering experiment, whose results were never released. Why would anyone talk about an experiment that hadn't produced results? And then there was the whole issue of the dental X-ray experiment. Carroll, a coauthor on the original paper, was making it pretty clear at that point that he didn't think that experiment proved triggering.

And that was the dilemma: either isomers were about to become the biggest thing in weapons research since nuclear fission, or the whole thing could blow up in Ullrich's face as an embarrassing dalliance into fringe science.

Ullrich was till trying to decide what to do about isomers when he ran into Bohdan Balko, a scientist at the Institute for Defense Analyses (IDA), a federally funded research center that specializes in military issues. Balko, like Ullrich, was a nuclear physicist, and knew Collins personally from back in the days of gamma-ray lasers and Star Wars; he offered some cautionary words to Ullrich. It might be best to check this whole thing out, he warned. At that point, Ullrich decided he needed a reality check. He asked IDA to evaluate the fantastic claims being made by Collins. In late 2002, three scientists at IDA—Balko, along with David Sparrow and James Silk—produced a ninety-eight-page report on the topic of isomer triggering. Unlike the JASONs, whose 1999 review of the subject lasted just one day, IDA exhaustively researched hundreds of papers on the subject, including those by Collins.

While the IDA report concluded that research on isomers should go forward, it was critical of the focus on weapons, and even more critical of Collins's work.

"Don't force it into trying to be practical before the relevant background work is done and it becomes ready for 'prime time,'" the authors wrote.

One of the interesting things IDA noted was that the gamma-ray laser idea seemed to have been dropped by most groups, who were focusing instead on nuclear batteries or explosives. In an only slightly veiled criticism of Collins, IDA noted the habit of some groups to "ignore results that were obtained by others," the "tendency to 'rediscover' ideas and promote them as one's own," and finally the "tendency to overstate results."

In a personal blow to Collins, the authors also concluded his *Physical Review Letters* paper was "flawed and should not have passed peer review."

Reading the report, and reflecting on his own views, Ullrich decided he would certainly be willing to support a basic research program that might advance knowledge of isomers, but not a weapon.

"The way we had left it was that we should endorse a continued research program to better understand the physics of de-excitation. But there was no basis for forging ahead with weapons," Ullrich told me.

Ullrich had done what one might expect any good government official should do. He had looked at the problem to see if it posed a threat to national security, solicited views from the experts, and then used those views, together with his own scientific judgment, to make a decision on what course of action might best serve the national interest. In a sense, he had repeated some of the same things that Peter Zimmerman had done a few years earlier at the Arms Control and Disarmament Agency. But if Zimmerman was instinctively opposed to hafnium because of its threat to arms control,

Ullrich was intrigued by what isomers offered to U.S. weapons research. From two quite opposite points of view, both had come to the same conclusion. Hafnium was not likely to be a weapon, and there was certainly no threat of the United States being scooped by Russia on an isomer bomb.

That may have been the end of the isomer bomb were it not for one problem; Ullrich himself was getting ready to leave the Pentagon to take a position in private industry. By the time the IDA study came out, Ullrich had packed his bags and was on his way out the door, just as another part of the Pentagon was jumping on the hafnium bandwagon: DARPA.

HAFNIUM WENT TO the Pentagon by way of New Mexico, helped along by a cadre of believers in the Air Force. One of those, of course, was Forrest "Jack" Agee, the Air Force scientist in charge of funding basic physics. He was the man who, in 1999, started funding Collins, while also publishing with him.

In early 2004, I went to visit Agee at his office in Arlington, Virginia.

Standing in front of the nondescript building that housed the Air Force Office of Scientific Research, I stopped for a moment to take in the gray façade that showed little sign of military occupancy. Office workers shuttled in and out of the multistory building, and it wasn't until I arrived at the Air Force's floor that a halfhearted attempt at military security was on display. A sullen woman reading a copy of *People* shoved a red badge at me, barely glancing at my press credentials.

Agee, once described to me as the *eminence grise* behind isomers, smiled as I entered his office and extended his hand like a caretaker greeting a mourning relative on their way to buy a casket. It was the last time he smiled. With dark-tinted glasses and a dour demeanor, Agee did not seem like the type of military official to give interviews, and I was surprised, in fact, that he had agreed to speak to me at all. Maybe he was surprised, too, because as soon as we sat down at the small oval table in his office, he immediately looked uncomfortable. Seated at the table, I noticed that Agee had a corner office, but with the windows blocked at every angle by adjacent buildings, casting the room in a permanent gloomy haze.

To Agee's right sat a public affairs official, and to his left, a security officer, who as Agee explained, was there to make sure he didn't say anything classified.

What secrets could accidentally slip out, I wondered?

As Agee explained it to me, he was familiar with isomers from back in the days of Strategic Defense Initiative—the Reagan-era ballistic missile shield—when he was in the Army, and worked on experiments related to isomers. Later he moved over to the Air Force, working as a scientist in the service's research laboratory in New Mexico, where he met Carl Collins and Patrick McDaniel.

But Agee also traced his interest to isomers back to Eastern Europe.

"We became interested in this research because we were aware of it and it was of potential interest to the Air Force," Agee said. "And we started out with a conference in Predeal, Romania to get some background on what had gone on in Eastern Europe. After the Wall came down and the Soviet Union broke up, it was an opportunity to learn more about their programs."

When I asked him about the controversial nature of the work, particularly the scientific debate around Collins's hafnium triggering experiment, Agee frowned deeply. "I know that work is going on around the world in this area," he said. "We are familiar with a number of countries that are pursuing this."

Agee paused for a moment to clear his throat and glanced out the window with its plaintive view of the next building—

perhaps thinking about the legions of foreign countries that could be eavesdropping on our conversation about dreaded isomer weapons.

He cleared his throat again, and then continued: "It was a surprise that Japanese torpedoes worked in a shallow harbor in 1941. We were technologically surprised by that and with awesome impact. So, the fact that there are countries other than ours that are working on this, well, we better be able to know what this is about whether we ever find an application for it or not, in case others find that."

According to Agee, his job was to ensure that "technological surprise" did not occur, or in other words, that impoverished Eastern Europeans, for example, did not suddenly come up with a new isomer weapon.

"The penalties you pay for that can be substantial," he said.

I was struck that just about every government scientist I'd met had described their job as preventing "technological surprise," but something like the isomer weapon was only a threat if it worked, or had a reasonable chance of working, I pointed out. What about the JASONs review? An expert panel of scientists had essentially said the hafnium bomb couldn't work, or at least had about as much a chance of being a bomb as a jelly donut. Was there really any legitimate fear of isomer bombs raining down on the United States in the near-to-distant future?

Agee scoffed.

"We rely on more than just a few days' review by some panel—albeit populated by smart people," he said.

What did the Air Force rely on, I asked?

According to Agee, the Air Force required its researchers to publish in the literature and attend conferences to "expose their ideas and findings to critical discussion."

That discussion, it turns out, mostly involved extremely impoverished Eastern European scientists. Agee, through his office, had actually been instrumental in supporting much of the work around the world on isomers. There were, according to Agee, over twenty "micro-contracts" as he called them—small contracts for just a few thousand dollars each—that went to researchers in the former Soviet Union and Warsaw Pact countries, including Romania, Moldova, and Ukraine.

Probably others countries as well, he said, though he couldn't remember. "My memory is one of the strongest supports of our national security, because I can remember virtually nothing so I am no threat," he said, without cracking a smile.

But the Air Force—like Collins—seemed to dwell in its own world. Because the conferences that his beneficiaries presented at were not typically those visited by physicists in the American Physical Society, the main scientific association of the United States. Rather, the Air Force's isomer scientists were presenting at an odd assortment of conferences held primarily in Eastern Europe. And those isomer conferences he mentioned—sponsored by the Air Force—were attended almost entirely by beneficiaries of his office. In other words, the "reviews" were coming from scientists from extremely poor Eastern European countries who would likely have little reason to criticize their U.S. sponsors, let alone admit that isomers weren't going to revolutionize the world.

And Agee's claim that isomer research was some esoteric field known only to his cadre also seemed slightly out of whack with reality. In fact, while hafnium triggering was indeed a venture unique to Collins and his colleagues, isomers were hardly an area known only to his group. Along with the Livermore lab's research in the area, there was a wealth of nuclear physicists with knowledge of high-spin physics to which isomers belong.

But independent review, were it there, would have been perhaps uncomfortable. Agee not only funded the work on isomers, but also appeared as a coauthor on papers with

Collins and Carroll. That meant Agee was funding research he himself was working on; the people reviewing the work were other scientists that he was funding; and Agee was using those reviews to justify continued funding. It didn't exactly seem like a system of independent peer review—or at least not like that used anywhere else in science—but it was a model that I saw with increasing frequency in the military.

So why was the Air Force funding isomers? Agee gave me an answer that didn't really provide any information.

"If this research pans out and we were able to trigger the release of large amounts of energy, I think the Air Force would find many applications for that," he said.

"Can you give some examples?" I asked.

Agee's tinted glasses seemed to grow two shades darker at my question. He furrowed his brow slightly and cast a furtive glance out the viewless window. "Applications start to drift into classified matters," he said quietly. "We'd prefer not to do that."

However, Agee suggested that I get in touch with Hill Roberts, the scientist from SRS Technologies in Huntsville, Alabama. Roberts, Agee told me, knew a lot about isomers.

GOOGLE NEVER QUITE orders information in the way you expect it would. When I entered Hill Roberts name into the search engine along with the word "hafnium-178," rather than scientific papers, I found myself staring an intricately drawn picture of an isomer on a Web site belonging to an organization called Lord I Believe.

The diagram was of hafnium-178, and the caption and preceding slide noted that the design accuracy of the unique little isomer was over one in one billion, or in other words, hafnium-178 was proof of an intelligent designer. That certainly offered a new twist on the hafnium debate, I thought.

Roberts, it turns out, was a very active proponent of something called Christian Evidences—a movement that melds science with Christianity to produce what struck me as a sort of feel-good version of creationism. Roberts, the founder of Lord I Believe, gave seminars and sold videotapes on Christian evidences, while lamenting on his Web site how the "worldview of naturalism dominates my scientific community."

He also sold hafnium, or at least rented it.

Further reading on the Web site described how Roberts, who had been touring Eastern Europe to give lectures on isomer research by day, was at night providing evangelical presentations to Russian scientists using diagrams of hafnium as an example of divine perfection. Given the propensity for hafnium to attract true believers, Roberts's unique use of hafnium wasn't that surprising. Divine visions of the hafnium bomb seemed a perfect fit for a physicist from Huntsville, a city that has long balanced messianic religious predilections with a scientific community that was initially transplanted from Nazi-era Germany.

When I reached him at his office down in Huntsville, Alabama, Roberts was polite and friendly and happy to talk about hafnium. I asked Roberts about his "extracurricular" activities in Russia, and he explained that at some of the places where he had given lectures about hafnium, physicists had approached him to give a talk about Christianity, and he had merely obliged. His religious work, he acknowledged, was not without controversy.

"For whatever it's worth, the rancor that gets heaped on me from that direction is worse than anything I get from hafnium," he said.

I also learned immediately that the most important role Roberts had in the whole isomer business had to do with his lock on the hafnium market. "The stock I have at SRS and the stock that Carl has access to at Texas consists of essentially the world supply," he boasted. "If somebody wants a sample of this material, we have it."

The hafnium Roberts owned was leased to scientists conducting experiments, he said. It was pretty good business.

His supply of isomer was stored down at his lab in Huntsville in a tiny little vial—suspended in just a few millimeters of liquid. "We have a good lock on that door," he chuckled.

Roberts's explanation for how he came to work with Collins seemed to fit well with everyone else's story. He was interested in isomers, read about the possibility of isomer weapons, and then he met Carl Collins.

"We didn't cast about too long before we jumped on the idea that we might be able to get energy from triggered isomers," he said. "We asked around at Livermore, and up at [Ballistic Missile Defense Organization] BMDO, and quickly made connections with Carl and began to discuss ideas with him in the mid-1990s or so."

Roberts's current relationship with Collins seemed a bit strained, however. He pointed out that while Collins was the lead experimenter in the dental X-ray experiment, it was Roberts's company that had paid for the Eastern Europeans to travel to Texas, and it was Roberts that had arranged, through friends, to get the money for Collins, or rather Collins's wife, to buy that first batch of hafnium. And that perhaps, was where the problems began. Collins had locked that sample up in a drawer in Texas. Several times, Roberts brought up the issue of the hafnium ownership. "Coherent Technologies consists of his wife," Roberts retorted. "I don't know of their business arrangement, but that's the only product that they have and, in the past, they don't . . ." Roberts stopped and cleared his throat. "In the past Carl carefully husbanded the material to the extent that he would not allow it to be used anywhere else except in that lab."

Roberts, of course, was also a participant in the original dental X-ray experiment, and a coauthor on the subsequent paper. Did he believe that hafnium triggered?

"My situation is that I have participated in experiments

that I believe I have seen triggering in, and I've participated in experiments where I'm absolutely certain we did not see triggering," he said. "I am at a loss to explain what the difference is."

The scientific controversy didn't much matter to Roberts, anyhow. "I don't care whether I publish a paper or not, I'm interested in business," he explained truthfully.

And the business of isomers was looking up, particularly in a world concerned about terrorist threats. While Roberts had received some funding from NASA to look at using isomers, and particularly hafnium, as a way to propel a future rocket through outer space, the real market was with the Pentagon. Isomer weapons, Roberts suggested, would be an effective way to strike stores of chemical and biological materials. The intense gamma rays would be one of the most effective ways to destroy those contaminants, he explained. That same power also made it an extremely effective weapon against people, at least in theory.

What Roberts described was a nearly apocalyptic vision of a destructive force not unlike that unleashed on the Nazis after they opened the cover of the Ark in the Steven Spielberg film, *Raiders of the Lost Ark*. Like the skin that melts off the Nazis' faces, Roberts had his own graphic description of the effects of an isomer weapon on humans. "It's kind of horrific," he said.

"They just turn to goo."

SCIENTISTS, LIKE MOST people, would rather get something for nothing. But for most scientists, getting funding is an arduous process that involves putting together detailed proposals, submitting them to funding agencies, and hoping that from a limited pool of money, a small piece will go to support their work. A critical part of this process is the way in which proposals are typically selected. They are peer-reviewed by fellow scientists, often those with specific expertise or experience in the topic being proposed, but typically with at least a general knowledge of the subject. Scientists tend to recognize that the unpleasant and competitive process of peer review benefits everyone, because only the best and most credible projects get funded.

There are often complaints that peer review is biased, that reviewers will give poor marks to proposals that contradict their own work or are competitive with their own funding, or worse yet, that scientists tend to band together, giving money only to their friends and allies. Many of these complaints have elements of truth, but like Winston Churchill's famous statement on democracy being the worst possible form of government, except for all others, scientists recognize that peer review, with all its blemishes, is probably the best way to divide up funding.

Scientists may wait impatiently for months to see if the National Science Foundation, whose average grant is some-where around $100,000, will provide money to keep their research going, graduate students employed, and generally make their life meaningful. With the Pentagon and science, it's a completely different game. The military services often don't use peer review for individual proposals. Neither typically does DARPA, which in many cases, provides funding of half a million dollars and up with little more than the signature of one or two officials. DARPA program managers have the ability to start up a project, often ones involving millions of dollars; all they need is approval of the director.

In 2002, Martin Stickley, newly returned to the Pentagon's research and development agency, approached DARPA's director with the idea of funding hafnium. Prove that you can produce it, was Tether's response. The world's supply of hafnium was in the micrograms—hardly enough to produce a useful weapon—and if it couldn't be produced in greater quantities, it wasn't of much use to the Pentagon. Stickley agreed, and hatched a grand plan. He would form the Hafnium Isomer Production Panel, a group of experts that would determine the most economical way to produce enough isomer needed for a bomb.

There were already ideas for how to do this. Stickley's colleague and Collins's old friend, Patrick McDaniel, had his own scheme for how to produce hafnium. According to one former defense official, McDaniel showed up at his Pentagon office and announced he was working on creating his own hafnium supply. "He was running a separate operation with Oak Ridge," the former official told me. "He was going to corner the hafnium market."

Oak Ridge National Laboratory is a Department of Energy laboratory in Tennessee that, similar to Los Alamos, was established in 1942 as part of the Manhattan Project. Oak Ridge produced the plutonium and uranium for the first atomic bombs, although, like the other labs, it eventually branched out into other areas of scientific research. When the United States stopped producing new nuclear weapons, its isotope production and separation facilities declined in importance. Then, in April 1999, just a few months after the first dental X-ray experiment was published, the Department of Energy allowed a private company from Georgia, called Theragenics, to lease Oak Ridge's plasma separation facilities.

Oak Ridge's privatization drive was fortuitous timing for McDaniel, who was looking for someplace to produce hafnium.

"Are you out of your freaking mind?" the former Pentagon official recalled of his immediate reaction to hearing about the plans for hafnium production. Without any solid proof of triggering, the hafnium believers were set to produce an isomer whose world supply could be measured in micrograms.

"That was very crazy," the former official said. "Nothing had happened that we were ready for that."

Yet with no oversight, no review, and no questions, Stickley in 2002 provided an initial $7 million in funding that was sprinkled out to various universities, national labs, and private companies. The largest beneficiary—over $2 million—went to Oak Ridge to investigate production. At that point, the Air Force's was in its fourth year of funding isomer triggering, but DARPA had substantially more money. The sudden boost to hafnium left some government officials shaking their heads, however. In November 2002, a concerned scientist in the Air Force's London office—the same place Stickley used to work—wrote an alarmed letter to two scientists at the Los Alamos lab.

"Gentleman" began the letter from Alex Glass, a PhD scientist and an Air Force official in charge of laser work. "Both DARPA and the Air Force have active programs in induced emission of gamma radiation, and I am concerned that not all the evidence is being considered in evaluation of various concepts."

Glass wrote that he had heard something about an experiment that contradicted Collins's results, and he wanted to know if those results were published.

Jerry Wilhelmy, one of the original participants in the Argonne experiment, wrote back to Glass, explaining that indeed, his colleagues had published the results of the experiment, and that they had found no evidence of hafnium triggering. He and his colleagues were hoping that someone would appoint an independent panel that would evaluate the various hafnium results and make some sense of the controversy. Perhaps more astonishing for Wilhelmy, however, was when over six months later, he received another inquiry, this time from a physicist working for Lockheed Martin Corporation, the world's largest defense company. The physicist, James Rejcek, noted that defense industry was also interested in hafnium, but he, too, was skeptical about the proposed results. Rejceck forwarded to Wilhelmy the Pentagon's proposed budget for hafnium.

Wilhelmy was in shock.

It was almost $10 million for fiscal year 2004 and $20 million for fiscal year 2005. In addition to the $7 million Stickley had already handed out, that meant DARPA was preparing to spend upwards of $40 million on a basic science project—an amount of money almost unheard of for physics research.

Noting that the published counter-experiments found no triggering, the Lockheed scientist questioned the rationale for such a major commitment of money. "I agree that an independent panel ought to review what has been done," Rejcek wrote.

That review, however, never happened.

Instead, in early 2003, William Herrmannsfeldt, a physicist at the Stanford Linear Accelerator Center in California, got a call from Tom Ward, a former Department of Energy official. Already in his seventies, Herrmannsfeldt had over forty years experience working on accelerators at Stanford; an iconic photo of Herrmannsfeldt working at the Stanford accelerator featured in an old issue of *National Geographic* even made its way onto a commemorative U.S. postal stamp.

Ward knew Herrmannsfeldt from another type of production panel they had both served on; that panel had looked at ways to produce tritium, a key ingredient in nuclear weapons. Ward, who had since moved over to a private company called TechSource, explained that he was helping to organize something for DARPA called the Hafnium Isomer Production Panel, and he wanted Herrmannsfeldt to be on it. Since accelerators were one of the schemes under consideration for making hafnium isomer, Herrmannsfeldt seemed a logical choice. That didn't mean he knew anything about hafnium, however.

"What's hafnium?" asked Herrmannsfeldt.

When Herrmannsfeldt was first appointed to the panel, he began his research into the obscure hafnium isomer the same way most people unfamiliar with the topic might begin—he typed the word into the Google search engine. His Web search led him immediately to an article in the *San Francisco Chronicle* about scientists at the national laboratories who had attempted unsuccessfully to repeat the Texas experiment. Then he found out about two Pentagon-backed studies that had panned the Texas physicist's work. It didn't sound good.

The Stanford scientist eventually called up Peter Zimmerman, by then a scientific advisor to the Senate Foreign Relations Committee, to ask him about the JASON study he had commissioned on hafnium. Sitting in his Senate office in Washington, Zimmerman was a little surprised to get a call from the West Coast scientist, but even more surprised when he heard that the isomer bomb was alive and kicking. He thought he'd killed it just four years prior.

In February 2003, Herrmannsfeldt made the short forty-mile trip to the Livermore lab to meet with some of the scientists involved in the 2001 experiment. What he found bothered him, the ominous shadow of cold fusion creeping in through the crack of a Pentagon door. All the elements were there: the incredible claims, the immediate doubt, and most important, the inability of independent researchers to replicate the results of the original experiment.

Trying to find ways to mass-produce isomers without even knowing whether it could be used in a bomb seemed downright insane to Herrmannsfeldt. The closest analogy that he could think of to isomer production would be antimatter, the *Star Trek*–reminiscent notion of a mirror-image material that explodes when it comes into contact with matter. Antimatter really exists—particle physicists had succeeded over the past few years in producing small amounts of anti-hydrogen—hydrogen with a positive charge. Also like isomers, antimatter tends to evoke images of *Star Wars* death beams, which would possibly be true, if the world's supply of antimatter consisted of more than a few atoms. The problem was that antimatter is ridiculously expensive to produce. Herrmannsfeldt wrote to a former colleague of his, Gerald Jackson, who was running a private antimatter business.

Jackson quickly replied to Herrmannsfeldt with the latest news from the world of antimatter. "It is funny that you should write me about this subject, since I have in my own way weighed in on it from a completely different direction: the use of the isomer for NASA advanced propulsion," Jackson wrote, noting that Hill Roberts, the hafnium believer from Alabama, had presented at a recent meeting on that very subject.

"The issues of experimental reproducibility and quantity both came up," Jackson said. "In fact, it was mentioned that antiprotons are more plentiful than the isomer!!!"

Jackson's company, Hbar, was looking at antimatter to fuel spaceships—futuristic certainly, but perhaps not as farfetched as a superbomb. In Jackson's view, antimatter was in fact much better than isomers. "The beauty of antiprotons is that technology for storage densities required for explosives are generations away," Jackson wrote Herrmannsfeldt.

Jackson's cheerful optimism on the future of antimatter wasn't helping Herrmannsfeldt much with isomers, or more precisely, his effort to get DARPA to take a long, hard scientific look at hafnium triggering before diving right into mass production. When Herrmannsfeldt suggested to DARPA, however, that the members of the Argonne group at least be invited to the production panel meeting to present their opposing results, he was firmly rebuffed.

Herrmannsfeldt had a less than fruitful discussion with Martin Stickley, the DARPA program manager, requesting that some of the scientific critics of the isomer project be invited to the meeting. Stickley reminded the Stanford scientist that as a participant in the production panel, it was not his place to say who should or should not be invited to its meetings. Herrmannsfeldt disagreed and his gripe was simple. He had been recommended to the panel because of his expertise in accelerators, but he was also educated as a nuclear physicist, and isomer triggering flew in the face of known physics.

Frustrated by the response and convinced that DARPA was on its way to the next cold fusion, Herrmannsfeldt decided he had to find a way to kill the project. And with that idea in mind, he launched what he later called a "guerilla warfare" campaign against DARPA's isomer program. He began by gathering signatures for a letter requesting that DARPA support an independent review of hafnium triggering before jumpstarting a program to build a weapon. Herrmannsfeldt circulated the draft letter around to various members of the production panel, to scientists familiar with isomer research, as well as to Collins's colleagues and critics.

Patrick McDaniel, upon receiving the letter, sent back an accusatory note, comparing Herrmannsfeldt's campaign against the hafnium bomb to the persecution of Galileo. In his response, McDaniel told Herrmannsfeldt that while he, too, had once been a skeptic, he was now "convinced" that hafnium triggers. "However, your letter propagates what I have come to realize as one of the most intense vendettas I have ever seen attempted by a considered educated individual, Mort Weiss, against Carl Collins," he wrote.

Why McDaniel chose to single out Weiss was unclear, but it followed a pattern long-since set by Collins—grandiose claims of persecution combined with personal attacks on the perceived instigators. But Herrmannsfeldt's letter never once attacked those results or Collins—it only asked for a review.

On August 13, 2003, Herrmannsfeldt sent his final letter

directly to Martin Stickley at DARPA and his colleague, Ehsan Khan, at the Department of Energy. The letter, which began with "Dear Friends," was signed by fifteen scientists, all urging an independent review of Collins's experimental results. The letter was signed by some of the nation's best-known physicists, including Wolfgang Panofsky, the founder of the Stanford Linear Accelerator Center. Even five members of DARPA's own hafnium production panel signed on to the letter.

"Faced with the controversy surrounding the history of the hafnium isomer concept, and with the numerous objections raised by the JASONs and others over any projected use of the hafnium isomer, we feel justified in requesting that photon-triggered deexcitation of the hafnium 178 isomer be placed on firm scientific grounds before proceeding to study applications that may not make physical sense," Herrmannsfeldt wrote in the letter. "A first step would be an independent expert review of the published experimental results."

Khan and Stickley were not pleased. They immediately called up the director of the Stanford Linear Accelerator Center, where Herrmannsfeldt worked, demanding to know if the lab intended to pay for the costs of such a review. Herrmannsfeldt reissued the letter, this time on private stationary, which led to another unpleasant exchange with Stickley.

"Martin basically told me I was crazy," Herrmannsfeldt said.

"What problem?" was essentially the DARPA response—according to Stickley, the triggering experiments were great and hafnium was right on track.

Herrmannsfeldt, as it turns out, was by no means alone in his concerns about the program.

James Decker, a senior official in the Department of Energy's science division, was getting complaints from all over the nuclear weapons laboratories about hafnium and considered pulling the department's laboratories out of the DARPA program. Stickley preempted the move, by writing directly to Decker, warning him that such a move, "would destroy the program" and would create a potential backlash. Even though DARPA was paying for hafnium, Stickley needed the support of Energy Department officials, like Paul Robinson, to have credibility. DARPA was also relying on Energy Department facilities and personnel to advise it on hafnium production. Without the Department of Energy, the hafnium program would effectively implode.

Pulling out of the hafnium program "would set back relations between DoE and DoD in a major way," Stickley wrote. "I am sure that you do not want to do that."

In addition to the overt threat, Stickley added a financial incentive to Decker for allowing the Department of Energy to stay in the program. "What you may not realize is that DARPA is funding about $4M worth of work in the DoE in addition to funding the HIP Panel," he wrote.

Stickley went on to exhort Collins's results, claiming the "world" accepted that tantulum-180 could be triggered. In a style that echoed Collins's pronouncements, Stickley added, "this is the existence of proof of isomer triggering." Stickley also noted that Collins's results were published in peer-reviewed journals.

Of course, in 2003, when Stickley was writing his letter to Decker, only a lone Russian publication was willing to publish Collins's results.

The members of Argonne group were not invited to the first hafnium meeting, Stickley explained, because they would be "disruptive." How they would be disruptive, Stickley didn't say, but he did write that withdrawing the Department of Energy's support "would create major problems that I am sure you do not want to bring on the DoE and DoD/DARPA." Decker, perhaps sensing he had a political battle on his hands, allowed the Department of Energy to continue working on hafnium.

In 2005, I asked James Decker's spokesman why the Department of Energy's Office of Science hadn't pulled out of the program, as Decker had apparently considered doing. Was it Stickley's threat to Decker of a Pentagon reprisal against the Department of Energy? Was it the lure of $4 million that Stickley promised Decker for the Energy Department's labs?

All I got was silence.

RARELY DOES THE public get a glimpse inside the closed-door meetings of official Washington, where this country's national defense is presumably debated and determined by elite officials using the best possible technical expertise. Had the public gotten a chance to peak inside a DARPA meeting held in the spring of 2003 on the subject of hafnium production, they would had a window into an alternate universe: a group of defense officials outlining plans to spend billions of dollars to mass-produce one of the world's most expensive materials to put in a bomb that could never exist.

The first meeting of the Hafnium Isomer Production Panel took place in May of 2003 at the northern Virginia offices of Booz Allen Hamilton—one of the Pentagon's largest contractors, which provided the administrative support for what was expected to be the beginning of a potentially large and lucrative defense project. Martin Stickley, DARPA's point man for the hafnium bomb, was on hand to brief a group of several dozen scientific experts assembled from across the United States' top laboratories and universities. A long briefing sent to scientists even prior to the meeting warned them that the entire project—dubbed the Stimulated Isomer Energy Release—was covered by the U.S. State Department's arms export regulations. Violations of those restrictions could spell fines or even jail time.

DARPA was taking the hafnium bomb with utmost seriousness. Knowledge about isomers and isomer production was to be strictly controlled, it warned. The accompanying briefing cautioned the scientists that "penalties are severe" for violating the restrictions, "and it is THEIR responsibility to: Ensure no export controlled material is released; ensure all documentation presented MUST be submitted through government public release process prior to public dissemination; and ensure Office of Defense Trade Controls, Department of Energy, and the Nuclear Regulatory Commission certifications are in place."

The scientists had been handpicked based on their knowledge of accelerators, separators, and reactors: all possible production sources of the hafnium isomer. The vast majority, however, knew almost nothing about the complex area of nuclear isomers, and many had never heard of isomer triggering before getting the call from DARPA. It didn't matter. Their mission, Stickley told them, was to help the U.S. government figure out a way to economically produce the isomer in usable quantities.

Given that current costs were running about a $1 billion a gram, a 100-gram hafnium bomb would cost the government a staggering $100 billion—high even by Pentagon standards. Cheaper production methods were going to be crucial to the project's success. Therein was the problem. The isomer of hafnium-178 does not occur naturally and scientists had been able to produce only small quantities, micrograms actually, of the material for use in experiments. The cost of running facilities to produce even the smallest amounts of hafnium was exorbitantly high. Even the microscopic amounts used in experiments sold for well over $100,000 a sample. The techniques used to produce small samples involved bombarding heavy metals with neutrons in a nuclear reactor, a method not economically feasible for mass production. To test and build a weapon, the Pentagon needed a cheap and reliable source of the isomer. The expert panel was to advise the Defense Department on how to do this.

The panel would be split up into three groups and given a year to research the issue and come up with the most economical production methods. The near-term goal, they were told, was to produce enough of the isomer material to use in an experiment that would demonstrate a bomb. In less than four years, Stickley expected to demonstrate a working prototype of a hafnium bomb, representing an entirely new class of weapons.

In other words, if claims about hafnium proved true, the panel of experts assembled that day in Virginia could be the beginning of a new Manhattan Project.

And while the experts assembled there may not have known much about nuclear isomers, much less hafnium, Stickley provided them with what he felt was the necessary background. With a colorfully presented and detail-rich PowerPoint slide briefing. Stickley told the assembled scientists about DARPA's isomer program and explained why the U.S. military was interested in producing the rare material. On the very first slide, Stickley showed his cut-away picture of the hand grenade.

The miniature bomb elicited some shock from the audience.

The two-kiloton bomb was a clear jab at legislative restrictions that prevented the military from developing so-called mini-nukes, nuclear weapons with an explosive yield under five kilotons. The point was not lost on Stickley, who explained on a colored scale that an isomer weapon would have an explosive power that fell just short of a nuclear bomb. Equally important, the explosive power was based on electromagnetic changes to the nucleus of the atom, and not splitting the atom. In other words, it wasn't nuclear fission, and thus wouldn't count as a nuclear weapon.

The status of the technology, Stickley assured the

members, was relatively advanced. There was already demonstrated evidence of triggering using low-dose X-rays, he claimed, and the next step was to figure out a way to produce the isomer in mass quantities. Production was the only major barrier to developing "the bomb concept," Stickley explained.

Of course, none of the counter-experiments were discussed that day.

The potential military applications for the isomer extended beyond a cartoon bomb. Reminiscent of Henry Kissinger's battlefield nukes, the isomer explosive could be packed inside artillery shells for an extra kick, or it might be used as part of the Pentagon's revived plans to deploy a missile defense shield—a scaled back version of the 1980s-era Reagan vision of Star Wars. An isomer bomb could be launched to intercept incoming ballistic missiles. Unlike the Pentagon's "hit to kill" warheads that rely on a "bullet hitting a bullet"—the head-on interception of an incoming missile—an isomer-tipped interceptor would only have to get in the general vicinity of an incoming warhead. The explosive power of the gamma rays would simply blow the missile out of the sky. To illustrate the point, Stickley clicked to another slide showing a hafnium weapon shooting out to intercept in an incoming missile heading for a city framed by high-rises.

Another possible application would be to use the isomer as an intense radiation device that could destroy storehouses of biological or chemical agents. This idea was particularly important because Pentagon officials had been arguing for over a decade that some of the toughest biological agents—such as the resilient anthrax bacteria—could only be destroyed by radiation released from a nuclear device. But dropping a nuclear weapon to take out a suspected anthrax cache might be a little unrealistic, not to mention poor public relations. In this case, the isomer bomb might be an acceptable alternative to a nuclear weapon.

Sitting in the room that day were a handful of people whose work over the years was closely tied to the project and instrumental in hafnium's meteoric rise in the Pentagon, including Nancy Ries, the veteran program manager from Sandia, who had led internal lab work on the isomer project since the mid-1990s; Patrick McDaniel, one of the original advocates of an isomer weapon; and Hill Roberts, the founder of Lord I Believe and the nation's only private supplier of hafnium.

Heading the production panel was Paul Robinson, the head of the Sandia lab. As the most high-profile official involved with the isomer program, Robinson was supposed to keep what was already becoming a contentious project on track. Robinson, the head of a nuclear weapons lab, also had good reason to keep the isomer bomb going. His lab was getting a sizeable portion of DARPA funding to work on isomers, with the possibility of even more funding should isomers prove to be as big a discovery as some would hope.

Indeed, in an internal newsletter published in March 2003, one Sandia official noted how important isomers could end up being for the nuclear weapons labs. "We really think there is a big opportunity there," James Tegnelia said. "If we do well with this DARPA activity, if we find some new things, that could be really significant for the Labs."

The cochair of the production panel was Ehsan Khan, a Department of Energy official assigned to DARPA to help run the hafnium program. Khan, who had earned a doctorate in nuclear engineering from the Massachusetts Institute of Technology, was the Department of Energy's point man for "alternative energy." He was known for providing a sympathetic ear to offbeat scientists that showed up at the department with occasionally bizarre proposals for zero-point energy, a concept that involves drawing energy from a quantum vacuum. Khan, at least in that sense, was the right man for the job.

George Merkel, another military official, presented the

Army's view of isomers, which included a possible weapon, but also a compact, clean energy source. Merkel acknowledged isomers were a "high-risk" venture that should proceed along scientific grounds, but it was clear that many of the people in the room that day already had visions of a full-scale hafnium bomb project. Edward Dron, a Sandia scientist, described in his presentation, a possible "Highest National Defense Priority National Defense Mission," as producing two hundred grams of hafnium per year for five years, "enough for 100 missile warheads."

But the star of that spring meeting was Carl Collins, the Texas-based physicist whose controversial 1998 experiment was the basis for the Pentagon's interest in the isomer. After the end of Star Wars, it had taken Collins ten years to find his way back into the Pentagon's spotlight. Collins was in top form that day as provided his own overview of progress on isomer research, hailing the remarkable successes his research had demonstrated.

"It is easy to release the stored energy," he declared.

Collins went on in his presentation to attack the work of the Argonne group, referring to their experiments as "published failures," and saying that the results were "without merit." Of course, none of the scientists who had either criticized Collins's work, or who had performed the counter-experiment, were present, and no one at the meeting seemed interested in discussing those experiments, or the two government studies that had reviewed hafnium triggering.

William Herrmannsfeldt, the physicist from the Stanford Linear Accelerator Center, was in the back of the room, sitting tight-lipped as a long succession of government officials and consultants provided their slide presentations. Herrmannsfeldt got a healthy dose of what those in the Pentagon jokingly call "death by PowerPoint"—where military officials compete for the flashiest presentations that include exploding bombs, embedded video clips, and computer animation.

He watched incredulously as Stickley gave his presentation. The last time Herrmannsfeldt saw Stickley was in 1979 and it was the day Stickley was being fired from his position at the Department of Energy. Herrmannsfeldt felt sorry for him then; he was with a small group of scientists giving presentations on heavy ion fusion at the Department of Energy, and Stickley was the head of the Inertial Fusion Office, an important position within the agency. In the middle of presentations, Stickley was called away. When Stickley returned, he announced that he had been fired from the office.

It was a near comical scene, Herrmannsfeldt later recounted. Stickley's staff sat glassy-eyed and stone-faced, while Stickley continued on asking questions, as if nothing had happened. Perhaps it wasn't even Stickley's fault he was fired, Herrmannsfeldt suggested. That office had long been a political tug of war, but there were other scientists who believed that even then, Stickley had a reputation for being on the wrong side of scientific debates. In either case, it was twenty years until Herrmannsfeldt and Stickley crossed paths again, this time on the hafnium isomer production panel.

And then there were the rest of the attendees. Herrmannsfeldt knew a number of them from his decades of work in the Department of Energy, but the mix of private industry and government agencies struck him as suspicious. It was hard to tell who worked for whom and how the various people were connected to different parts of the project, except that everyone was getting money. "The Brookhaven Technology Group" had nothing to do officially with the Brookhaven lab, a Department of Energy facility, and consultants from private companies with names like "TechSource" crowded the room. All of them were getting a share of the approximately $7 million that was being handed out that year, with more money—in the tens of millions—expected in the near future.

No one at the meeting had bothered to mention a few of

the things that Herrmannsfeldt learned from his early calls, for example, how independent researchers—in fact some the country's leading experimental physicists—had been unable to replicate Collins's astounding hafnium results. Also, no one mentioned that two independently commissioned government reviews had said the science was wrong.

When Herrmannsfeldt after the meeting called Peter Zimmerman, who had commissioned the original JASON review, Zimmerman was also shocked.

"Many presentations depicted the goal of the project as the fielding of a new nuclear weapon," Zimmerman wrote after reviewing the presentations. "This is contrary to the policies of the present administration and the previous two administrations. Because the yield of such weapons would certainly be less than 5 kilotons, work on their development is forbidden by statute."

Another problem, according to Zimmerman was that an "an isomeric explosive device would fall outside of the nuclear weapon design space explored over the last 40 years, a full-yield operational test would be required."

That could potentially mean that the isomer weapon could roll back a decade-old moratorium on nuclear testing.

But the question was: What to do? The $7 million going to hafnium—though big money within the scientific community—was hardly going to evoke much outrage in an annual defense tab that topped $400 billion a year, particularly with national attention focused on the war in Iraq.

When, in late 2003, I called the Federation of American Scientists, a Washington think tank founded by veterans of the Manhattan Project, I got a rather succinct explanation for why the hafnium bomb was not getting more attention.

"I think the critics recognize that by Department of Defense standards, it's not a lot of money," said Ivan Oelrich, a former government scientist working for the federation. "Even if they think it's a total waste, why lose sleep over it?

The Defense Department spends about $16,000 a second, so by DOD standards, it's not much to worry about. That might be part of the explanation."

But money alone should not be the standard for judging hafnium, he said.

Oelrich was familiar with isomer research from his days at the Institute for Defense Analyses, the same government-supported research center that had reviewed Collins's hafnium results in 2002. According to Oelrich, he had looked at models for the gamma-ray laser using isomers some fifteen years ago, when support was coming from various parts of the Pentagon. "In a very simple model, it isn't ridiculous on the face of it," he said. "It just doesn't work."

Though DARPA reasonably wanted to err on the side of pushing technology too hard, rather than being too conservative, "there have to be some standards," he said.

Triggering hafnium, in his opinion, just didn't meet any intelligent standard.

"You have to have some point beyond which it doesn't make sense to invest," he said. "Isomers should have reached at this point."

After all, by 2003, the idea of isomers triggering had been around for at least a couple decades. "It's been around so long, that's what is so unusual about it," Oelrich told me. "It dies, and then it resurrects itself. You would think that after so long, after the gamma-ray laser, we'd put it to rest."

But in 2003, with DARPA ramping up to start a full-scale hafnium bomb project, putting isomers to rest wasn't even being considered.

9

ISOMERS HIT PRIME TIME

THE CAPITAL BELTWAY is a world away from Collins's lab. Unlike the sparse landscape of the University of Texas at Dallas campus in suburban Richardson, Texas, northern Virginia is dotted with hotels where defense contractors, scientists, researchers, and "Beltway bandits" come to visit with their Washington sponsors. Common perception holds that the Pentagon itself houses the Defense Department, but in reality its offices extend along the Metro's Blue Line, from Rosslyn, Virginia, where acquisition managers line Wilson Boulevard to Crystal City, where satellite offices handle everything from foreign military sales to the management of a $200 billion fighter aircraft program.

The world of defense contracting blends seamlessly into the military tapestry of the Washington area. In many cases, private contractors work in Pentagon offices providing "technical assistance" and the near-imperceptible line between industry and government disappears in the subtle shades of color-coded ID badges: pink for contractor, white for Defense Department. Long before anyone fretted about the private warriors in Iraq, contractors were silently populating the Pentagon's offices without anyone blinking an eye. And for every government official who steps quietly through the "revolving door" to industry, there is a private employee stepping through the other side back into government. Military officials "retire" from government, but never leave their desk, rehired in the same position as a private contractor, often at twice the pay. In many cases it is not the door that revolves, but merely the payroll slip coming in at the end of the month.

And then there are the contractors from other cities who congregate in Washington to meet with their military sponsors. Visiting defense contractors meet in conferences, on panels, and at seminars in the Sheratons, Hiltons, and Marriotts that populate northern Virginia. While rarely five-star accommodations, they are all a convenient ten-minute drive to the Pentagon. The Hilton Towers in Ballston was definitely not for the high-end bandit. Sandwiched between a Metro station and an office building, the lobby of the hotel was the size of a typical family dining room, and the industrial carpet already showed the wear of daily visitors rushing to early-morning appointments.

Sitting in the lobby in early November 2003, an Army captain read the sports page of *USA Today*, with a PowerPoint briefing at his side marked "Transformation, Now!"—the Pentagon buzzword of the day. Rising from the opposite couch, a woman enthusiastically greeted a Russian doctor. They were off to a biological and chemical defense meeting. And there, too, was Carl Collins, dressed as you might imagine any professor, with a golf shirt and sports jacket. Tall and silver-haired, the Texas scientist was in his sixties, but he had the animated dynamism of someone much younger. It was my first meeting with Collins, who was in the area to brief DARPA's higher-ups on the progress in hafnium triggering, yet he seemed somehow out of place in the goldfish bowl of northern Virginia.

The DARPA isomer bomb project was already in full swing, with plans to spend up to $30 million on it over the next couple of years—all based on Collins's 1998 experiment. That program, and its funding, was also at the center of a raging scientific controversy. But if the ongoing debate bothered Collins, he certainly didn't show it that day. He was all smiles.

We debated for a minute the choices for breakfast and settled on the hotel's gloomy restaurant upstairs, which sported a sad buffet of curdling eggs and soggy bacon. Collins, a relatively thin man, said he was on the Atkins diet and selected the bacon. Later, when I would insist on paying, Collins would apologize profusely for choosing the "pricier" restaurant over the small take-out café downstairs. Gracious to the end.

Defense contractors call the Pentagon "the customer," and speak about program "milestones," mixing military metaphors with business euphemisms and indecipherable acronyms. Soldiers are "warfighters," a cancelled contract is "defunded," and an engineering challenge is a "long pole in the tent." The typical defense contractor is a potbellied former military officer still sporting the buzz cut and engineer's glasses, a throwback to the 1980s heyday of military aerospace.

Collins, on the other hand, did not walk the walk or talk the talk of a defense contractor. He spoke about the scientific method. He said he never had a security clearance, and didn't want one. He preferred doing research out in the open and wanted to continue working with colleagues from behind the former Iron Curtain. Although not particularly bothered by the military applications that had caught the eye of DARPA, he seemed genuinely uninterested in its focus on weapons. He said he really wasn't aware of how or why the agency became involved in his research. He had a contract in place with the Air Force, and at some point he simply noticed that part of the money was coming from DARPA.

Unaffected and exceedingly polite, Collins had the sort of chatty openness that made him instantly likeable. He spoke with a deep Texas accent—almost a self-parody in its authenticity—and he could shift effortlessly from discussions of quantum mechanics to his favorite local barbeque joint. Collins enjoyed talking about his work and heaped praise on his colleagues, but also didn't feel the need to parse words when it came to his detractors. He described his critics as "opponents," "enemies," and "politicians," who were scared by his work and what it represented—a challenge to the accepted laws of physics and a possible gateway to the next superbomb.

Speaking about the Argonne group—the collaboration of fifteen scientists across three premiere national labs that had disputed his results—Collins good-naturedly explained that he was simply mystified by their actions. No one in the group went to the professional meetings that he attended and none of them even tried to contact him about the experiments, Collins claimed. And then there were a few who were not professional at all, Collins said, declining to name names.

"One of the ones out of fifteen is a hero," Collins exclaimed, declining to mention the scientist's name for fear of causing him problems. Collins paused, and noticing my empty cup, quickly moved to pour me more coffee, and then continued. "We started talking together and we found out there are two aspects why this experiment might not work," he said.

According to Collins, the Argonne group had used the wrong X-ray levels, and even worse, didn't look at the right emissions—evidence of triggering. "The 130 fingerprint, the KeV line–that's in the article," Collins continued. "It is not normally there, but when you trigger, or speed up the decay, you get the emission of that gamma line. That's one of the signatures that you got something," Collins said.

Heck if I knew what a "130-KeV line" was, but I liked listening to Collins. It sounded like an almost reasonable explanation to me, or at least a genuine issue for debate.

The "hero" of the group had conceded that the Argonne detectors couldn't see that, according to Collins. "Then a very curious thing happened," Collins said. "Someone said, 'We've got to quit communicating among ourselves and have a review panel.'"

And that was the last he heard from the hero. "But he knows they've conceded at this 130 line," Collins concluded.

Before the whole issue was politicized, isomer triggering used to be a lot of fun, Collins lamented. Now his enemies were trying to stab him in the back, cut off funds, and have everything determined by panels. His opponents were writing secretly to the scientific journals, denouncing his work as a fraud. "I think there is a big fundamental problem," Collins said. "They are from the most expensive lab in the world and a bunch of tattered academics from a lot of different universities and countries got together and found something that the most expensive lab had the responsibility to look for, but couldn't find."

And then, with a tad of sympathy, he added, "We're all getting older, and a lot of great names couldn't find [triggering]."

Collins was bothered by some of the bad press his work was getting, declaring a recent *San Francisco Chronicle* article on his work a "hatchet job." The article focused mainly on William Herrmannsfeldt's campaign against the hafnium bomb.

The JASON review panel was also something that bothered Collins, who had a flair for drama. He saw himself and his challenge to traditional nuclear physics as the modern-day equivalent of the trials of Giordano Bruno, the Dominican monk who was burned to death in 1600 for claiming earth revolved around the sun. The "expert panels" represented by the JASONs and other critics were using the same scare tactics, according to Collins. "You start talking about expert panels, that's exactly what they did to Bruno," Collins said. "This is the same thing."

Collins also had a sense of humor, and laughed at the mention of the now-infamous dental X-ray machine.

"That's not the worst of it," he said with a grin, leaning over for emphasis. "Sometimes we used car parts."

There was nothing wrong with using a dental X-ray machine to save money, he said. And more importantly, Collins noted that in 2001 he and his team, working with Japanese colleagues, went on to validate the original results at the world's most advanced X-ray source, the SPring-8 synchrotron near Osaka, Japan. Collins chose SPring-8 precisely because he needed an advanced synchrotron, which can be tuned to precise energy levels. In his mind, this was the scientific method at work, the replication of earlier results proving out his theory of isomer triggering.

Collins acknowledged that in science, it was very hard to prove something is wrong, only that it's improbable. But Collins suggested that isomer triggering could be reduced to what he called an "exclusive or," or XOR, a logic formulation often used in computer equations. Write down two mutually exclusive statements, Collins said, and if you can show that one is false then the other has to be true. "One statement is everything we've seen and reported can be explained by something that does not involve isomer triggering," he said. And the "or" statement would be, "One thing we've seen cannot be explained away," he added. "

"OK, now, if one thing cannot be explained away, in a sense, you've proved there's an effect. You may not know everything about it, but you need to start learning about it. That's your two choices," he said, echoing the old Sherlock Holmes theory of proof. ("How often have I said to you that when you have eliminated the impossible, whatever remains, however improbable, must be the truth.")

Collins said his group has also conducted experiments at another advanced X-ray source—in Europe—but he declined to name the location, saying that he worried with all the recent press that it would cause his colleagues problems. Some people were concerned about the weapons issue, Collins conceded.

On that fall day in 2003, Collins said he was unsure that hafnium would be useful for a bomb, though he claimed not to have given it much thought. If he had given any thought to applications, he said, it was to the concept of using tiny amounts of hafnium "seeds" for cancer therapy. The isomer seed, Collins said, could be "triggered" to give out precisely the right amount of gamma rays needed to destroy a tumor. There was already some interest in this from the Mayo Clinic, he said.

But what if hafnium does end up as a weapon, I asked?

Collins placed his coffee cup firmly down on the table and spoke solemnly. "What is our job as scientists? To censor thought and development because we might be wrong?"

And then, in a theme that Collins would repeat to me many times over the next few months, he spoke about his view of the isomer weapon. "Poor Alfred Nobel, he invented dynamite; thought he was doing good things for society. He was so appalled at the end of his life that he left all his money for the Nobel Prize."

And what, I asked Collins, if isomers, like dynamite, ended up revolutionizing warfare, just like some in the Pentagon were claiming?

"These things happen," was Collins's response.

Collins paused to look at his watch. We had been talking—or more precisely Collins had been talking—for almost two hours, and he still had more to say. The hotel restaurant had grown quiet and we were sitting alone at a center table. Looking up, I realized a few Ethiopian waiters were leaning against a nearby pillar, listening intently to our curious conversation about science and bombs.

Collins had to pack his bags and catch a plane back to Texas. He got up and thanked me politely, remarking on what a pleasure it had been.

He rushed to the door. Turning back halfway, he stopped to say, "Even if you do a hatchet job, it sure was a pleasure." He smiled and waved goodbye.

It occurred to me that he really seemed to mean it.

And that was my first meeting with Carl Collins. Giordano Bruno confronting a scientific witch-hunt. Alfred Nobel sparking a revolution in weapons. With unnamed heroes he wanted to protect, enemies he couldn't name, and secret facilities that he didn't want to endanger. Collins was charming and gracious, and his version of events was so clear and compelling.

If only it was believable.

10

A BOMB AND A PRAYER

IF EVER THERE were a scientist who would have understood the nature of the hafnium bomb, it would surely have been Irving Langmuir. The famed chemist and physicist was known, among other things, for his common-sense approach to the physical world guided by a fascination with the way things work. That fascination, combined with a first-rate scientific mind, eventually led Langmuir in the early part of the twentieth century to study the ordinary light bulb. While he may have gone on to win the Nobel Prize, most Americans would likely remember him as the inventor of the gas-filled incandescent lamp, which still graces the ceiling of most homes to this day.

Langmuir, however, added one other major contribution to science.

On December 18, 1953, Langmuir gave a talk at the General Electric Company that left an indelible mark on the modern understanding of science. It was on the subject of "pathological science," or as he put it, "the science of things that aren't so." In his speech, saved from oblivion by a rough transcript found after his death, Langmuir laid out a number of examples of science gone awry, ranging from a supposed new form of radiation, called N-rays, to ESP and flying saucers. Unlike "pseudoscience," the attempt to dress up non-scientific ideas as science, or "junk science," which typically describes the use of shoddy techniques or poor methods to promulgate a desired scientific conclusion, pathological science is indeed a world unto itself.

In later years, long after Langmuir's death, the term *pathological science* was attached to any scientific pursuit where the scientists involved would claim that the inability of others to reproduce their results was caused by some fundamental flaw in others' work. Unlike junk science or pseudoscience, pathological science was often the product of credentialed scientists—sometimes even great scientists—who in finding a seemingly groundbreaking result were too quick to announce their discovery. And later, as other researchers failed to reproduce those results, the alleged discoverer would cling that much closer to the original claim, coming up with ever more elaborate reasons why others had failed. If indeed the major warning sign of pathological science is the scientist's claim that his or her experiment is so special, that others are somehow missing key steps that explain the lack of reproducibility, then Collins was marching happily down the trail blazed by the purveyors of N-rays and ESP.

In reality, Carl Collins was having some difficulties in the latter half of 2003 that should have been throwing up some definite warning signs. And even if Collins, who only ever referred to the agency as "the sponsors," really was oblivious to DARPA's support, the Pentagon's money also came with greater scrutiny. Along with William Herrmannsfeldt's ongoing campaign to kill the program, the scientific and political attacks against his work were also growing. What's more, Collins's cryptic version of events simply didn't match reality.

First, there was the matter of Collins's "hero," the brave scientist on the Argonne team who had conceded his group's detectors were "blind" to the 130 KeV line. The hero, as Collins would later admit, was Jerry Wilhelmy, a Los Alamos scientist, and he had a very different description of his brief encounter with Collins.

"Hmmm, I am not sure that I am flattered by being a 'hero' of Carl Collins," Wilhelmy remarked to me. "Actually I only sent Collins one e-mail."

Wilhelmy's one note to Collins was more or less by accident, the Los Alamos scientist explained, and was far short of the backdoor collegial conversation Collins had described. At the time, Wilhelmy was trying to lobby DARPA to appoint a panel to review the hafnium work, and copied Collins on the e-mail out of professional courtesy. Collins replied directly to the Los Alamos scientist. The brief exchange was not a pleasant experience for Wilhelmy, who after his run-in with the Russians, was again realizing how off-the-wall his encounters with isomer controversies could be.

Zeroing in on the "130-KeV line," Collins's first note to Wilhelmy insisted that the Argonne group "rectify the misstatement" that they could not see this new line. The Argonne detectors were blind to the new line, Collins asserted. "It is quite egregiously misleading the way you have stated it and an Errata is needed," Collins wrote to Wilhelmy.

The note was a bizarre one for Wilhelmy, in part, because it reflected that odd tendency of Collins to focus on details that were of no consequence to science. It was also an attack uncharacteristic of the usual polite discourse among physicists. In fact, there was no evidence that the "130-KeV line" was proof of triggering, nor was it even emphasized in Collins's earlier publications. That new line had more or less emerged like a virgin birth in a later publication, without any scientific reason why such a new line would exist.

Trying to move the conversation back to the science, Wilhelmy pointed out among the many known gamma-ray emission lines that the Argonne group measured, his group found no evidence of triggering. The hafnium isomer emits several well-established levels of gamma rays as it moves to the ground state, and Argonne had measured those lines very carefully, and saw no evidence of triggering. So, Wilhelmy wrote, unless Collins wanted to "compound the miracle of triggering," he had to explain why Argonne also didn't see triggering at any of the expected lines.

Collins wrote back, seizing again upon the magical 130-KeV line, and ironic for a scientist who employed a dental X-ray machine in his first experiment, accused the Argonne group of using outdated equipment. "I think that your Sponsors should purchase some better instrumentation for you so you will have a better chance of success in the future," Collins jeered. He didn't address the long list of scientific criticism that Wilhelmy pointed out in his e-mail. Instead, he again insisted that Argonne submit an errata to its *Physical Review Letters* article. Collins also rejected Wilhelmy's request to support an independent scientific review of hafnium triggering.

"As for independent confirmation of Hf-isomer triggering, that is how I think this all started," Collins sniped. "Another DOE Group of high esteem has done just that."

The other "DOE Group," was in fact, Patrick McDaniel, Collins's old friend from the Air Force Research Laboratory, who was performing experiments at a high-tech X-ray facility at Louisiana State University and at Air Force facilities in New Mexico.

Collins concluded his e-mail with a "tally" of the scientific battle:

> The score is now:
> YES: SPring-8, Paul Scherrer Inst., [Louisiana State University] CAMD, and Sandia;
> NO: [Argonne Advanced Photon Source] APL

Wilhelmy never wrote back.

And then, of course, there was Collins's chief enemy, Donald Gemmell. In fact, Gemmell had spent forty years as a physicist at the Argonne lab, ten as its director of physics,

and had a good sense of humor about the hafnium controversy. And Gemmell, who had so angered Collins with his
"cold fusion" accusation, had a reflective, if almost empathetic
approach to Collins at one point.

"Fascinating!" Gemmell wrote, when I informed him of
his chief enemy designation. "But you've made my day!"

Collins had depicted Gemmell as a sort of shady Machia-
vellian character, secretly contacting journal editors. Gemmell, when I contacted him, however, cheerfully turned over
his entire e-mail archives detailing his sometimes-convoluted
correspondence with Collins. It followed a typical pattern,
beginning with an initially polite exchange, followed by
Collins's rapidly escalating accusations of misdeeds.

In fact, Gemmell had first contacted Collins in March
1999, just a few months after the first publication of hafnium
triggering, to invite the Texas scientist to a conference to
discuss his results. Collins declined, saying he was too busy.
In July 2001, Gemmell made his second and last attempt to
communicate with Collins. He politely wrote him to explain
that his group would like to see Collins's latest results, and
requested a preprint of his upcoming article—a quite normal
request. Collins, in his reply, declined to provide the article,
and instead, launched into an attack on the Argonne group's
work. "I am telling you this at this time is that for another
30 days there remain to you 3 options: revise, errata, and
reply," he wrote. "After 20 August there will be only errata
and reply."

Gemmell was equally amused at the "130-KeV" line
controversy.

"The solid line is a 'fit' to the data points which look like
a random sprinkling of flies' droppings," he wrote me. "This
is the 'evidence' for the 'NEW 130-keV line'!"

Then, there was the "secret institute" where Collins was
performing his hafnium triggering experiments. In fact,
according to DARPA documents, and as Collins himself

would later concede, his team had conducted a triggering experiment at the Paul Scherer Institute in Switzerland. The managers at the institute in Switzerland—a country that prides itself on neutrality—grew upset at press coverage over a "superbomb," and quickly asked Collins to move his research elsewhere. When I asked Collins about the Swiss facility, Collins conceded the reason for his problems there had to do with weapons. "The managers are very upset with the recent press coverage, though no fault of ours," recalled Collins, who had just finished interviews with Radio Netherlands for a story on isomer bombs.

"They are in a bit of a sensitive position, understandably not wanting to be identified as doing bomb work for the U.S.," he wrote me. "However, their 'cover' may be 'blown' soon anyway."

But Collins said he saw no reason why that should affect his work. "In another week I guess they will have to consider realistic options, as opposed to ideals that can be dreamed by 'neutralists,' blessedly free from politics," he said of the Swiss, pointing out that the scientists there had agreed to attach their names to his upcoming publication. The article would go to print with their names, whether they liked it or not.

"Quickly, I say that we never misled anyone," Collins added, explaining that he had simply submitted his research proposal, never mentioning DARPA's interest in isomer weapons.

In fact, a few other problems emerged at the Swiss X-ray facility. Along with the concerns over weapons, which eventually ended Collins's experiment at the facility, there was the matter of outside observers. Under pressure to answer criticism, Martin Stickley hired a married couple, a pair of British physicists named Jirina and Nick Stone, to observe and review Collins's experimental procedures. Stickley expected that the independent observers from the prestigious Oxford

University would add credibility to the experiments. Apparently, his plan didn't quite work out.

As Nick Stone began to ask tough questions, Collins became more and more irate until the Texas scientist demanded the British couple leave, Stone later reported. Stone was less than impressed with Collins's work and began to do a bit of research on his own, eventually stumbling upon the work of James Carroll, Collins's former student. Carroll was surprised when the Oxford scientist called him out of the blue one day in 2003. Stone said he had heard that Carroll no longer supported his onetime mentor's work.

Stone wanted to know if Collins was an outright charlatan.

Carroll said he didn't like the idea of name-calling, but told Stone what he had explained many times before. He did not disavow the dental X-ray publication (after all, he was a coauthor on the paper), but simply now questioned whether it was clear proof of triggering. The Oxford professor listened carefully, and then told Carroll that he wanted to send him a copy of the report that he had written for DARPA, but Stickley wouldn't let him give it to anyone.

At least for the duration of the DARPA program, that was the last Carroll, or anyone else, heard from Nick Stone or about the mysterious report.

In the meantime, as word got back to the scientists who had participated in the Argonne experiment that Collins was calling them failures in public, things got nasty. The scientists—some of the nation's leading experimentalists— were not pleased by Collins's rhetoric. In an unusual move in scientific disputes, the general counsel of the Argonne National Laboratory wrote a warning letter to Collins shortly after the May meeting, calling his statements "inappropriate, especially given that Argonne was not allowed to participate in the panel."

The situation was even worse at Livermore, where nuclear

weapons scientists were getting what seemed like unreal reports of the "nuclear hand grenade." John Anderson, a former Livermore physicist who had known Collins from back in the Star Wars days, was apoplectic, and sat down to pen the beginnings of an editorial:

> Most Americans believe the Iraqis are the chief source of GI casualties today. In the future however, if DARPA program manager Martin Stickley has his way, we will start killing our own troops! In a recent presentation Stickley listed as a program objective the development of a five-pound hand grenade using the energy stored in a nuclear isomer. For about a dollar you can go to a grocery store and buy a five pound bag of flour, take it to your backyard an give it a hefty heave, How far did it go? Twenty feet? Thirty? Forty? Or even fifty feet? Stickley's hand grenade is supposed to explode with the force of two thousand tons of high explosive! Sorry, but I doubt even the dog tags survive.

The idea of a nuclear hand grenade was just about the stupidest idea Anderson had ever heard. A soldier would die of radiation almost as soon as he pulled the pin. "Some field weapon for our GIs!!" Anderson scoffed.

Seemingly oblivious to the Stones' presumably negative report and the mounting criticism from the Department of Energy laboratories, Stickley was oddly pleased with Collins's results and sent out e-mails to the critics on July 3, announcing the successful results of the latest experiment in Switzerland. Collins's new results, Stickley wrote, also helped explain why the critics had failed to see triggering. The critics weren't taking the measurements at the correct time. Stickley knew that to be true because Collins told him so, and that seemed to be Stickley's evidence for just about everything. Stickley was convinced Collins's results were correct. In one mail sent to a number of Collins's critics following experiments done in 2003, Stickley attached Collins's most recent publication from a Russian journal. "I urge those of you who continue to question his results, to please study this paper," Stickley wrote.

McDaniel, Collins's colleague, had also conducted successful triggering experiments, getting even better results than earlier experiments. Stickley repeated the claim that Patrick McDaniel had made in May at the production panel meeting—that the Air Force scientist's work had demonstrated an astounding 15 percent triggering. In other words, things were going great.

"Have a great 4th of July!" Stickley concluded.

WHILE STICKLEY WAS touting the successful results of McDaniel and Collins, he had somehow failed to mention that, in the meantime, James Carroll was repeating the same experiments, and striking out each time.

Carroll spent eight years working with Collins down in Texas, four as a graduate student and four more as a post-doctoral fellow. Relations between the two had long since broken off, but they were still the two scientists most closely associated with isomer triggering, and it didn't help that people would often confuse their close-sounding names. They somehow always ended up at the same conferences, speaking on the same subject, one after the other.

For the most part, Carroll tried to avoid confrontation, and was more worried about keeping up his own gamma-ray laser program at Youngstown University, a small campus in central Ohio. And he seemed to be doing a pretty good job. Carroll sent me a press release with his picture and an announcement that his funding from the Air Force, to the tune of almost half a million dollars, was the largest single contract the Midwestern university had ever received.

There were just two words that Carroll, once Collins's favorite student, couldn't stand to hear: cold fusion.

"Let me start by saying this is not cold fusion," Carroll said when I called him at his office. Those words were particularly stinging because Carroll had started graduate school in Utah, birthplace of the now-infamous cold fusion controversy. It was the year after he left University of Utah that two chemists at the university—Stanley Pons and Martin Fleischmann—made their astonishing claims about demonstrating room-temperature nuclear fusion. Carroll, meanwhile, had transferred to the University of Texas at Dallas, fascinated with the prospect of a gamma-ray laser, which struck him as an exciting opportunity to work on a possible scientific breakthrough. The Utah–cold fusion connection stuck with him, however.

"People for years still teased me. 'Were you involved in that?'" Carroll said of cold fusion. In fact, he wasn't, but the controversy shadowed him to isomer triggering, where at conference after conference Carroll would have to explain that his work was not at all a new cold fusion. Carroll said he didn't repudiate the results of the now-famous dental X-ray experiment, it just wasn't clear to him that it was clear proof of triggering. The results were "intriguing," but not conclusive, he said.

But a bigger problem was that Carroll's most recent experiments only seemed to contradict Collins's work, and the original Texas team had split in two parts. Sarkis Karamian, one of the Russian collaborators on the original dental X-ray experiment, had defected from Collins's group to work with Carroll, and Collins's team had dwindled to just a few members. Ironically, at one experiment in June 2003 at the SPring-8 facility in Japan, the two men actually crossed paths, although they barely exchanged words. Working at different parts of the facility, Collins claimed his experiment a success while Carroll said his group found nothing. And Carroll conducted another experiment at an advanced X-ray facility in Canada. Again he found nothing. And yet another experiment at the Brookhaven National Laboratory on Long Island also showed nothing. There were only so many times you could do null experiments, he pointed out.

"Why is hafnium-178 so sexy and why is there so much interest by DARPA?" Carroll asked rhetorically. Because, he explained, the hafnium isomer stores two and half million electron volts of energy, or 2.5 MeV, and if a small spark could trigger a release of that energy, it's a huge deal. Theoretically, that would make for a good bomb. But even if triggering works, scientists need to understand how it works, and the applications, if any, would be decades away, if not longer, he said.

The problem with the DARPA program, according to Carroll, was that it was focused on just one isomer, hafnium-178, and just one goal, low-energy triggering. Carroll described DARPA's involvement in isomer research as an "impedance mismatch," a scientific term describing the result of joining two systems that have different conceptual bases. DARPA wanted a fast track, he explained, but isomer research was still at the very basic stage. "It's a mismatch between expectations and reality. That perhaps is the difficult part here."

As for weapons, Carroll said he hadn't thought a whole lot about the applications and was more focused on the basic physics. When pushed, he admitted that a weapon was the most obvious application, though by no means the only one. "Maybe you can never make anything practical out of it, be it medical, industrial, or counter-terrorism. Maybe none of it will pan out. But in the meantime, there's a lot we can learn about how the nucleus responds to people banging on it."

Carroll said his group no longer planned to be involved

in the DARPA hafnium program once he completed the last of his experiments, set for Japan in mid-2004. After a series of results showing no indication of low-energy hafnium triggering, he wasn't interested in proposing any more hafnium experiments.

How much can you learn from getting nothing?

WHAT OF PATRICK McDANIEL'S miraculous results? McDaniel, who by 2003 had left the Air Force and was working at the Sandia lab, had reported getting 15 percent triggering, which if true, should have been convincing proof of triggering and a much stronger result than even Carl Collins had claimed. Following the 1998 dental X-ray experiment, McDaniel had conducted a string of experiments at the Air Force Research Laboratory's facilities at Kirtland Air Force Base in New Mexico and at an advanced X-ray source at Louisiana State University. But no one had ever been able to scrutinize those results because they had never been submitted for publication.

"The problem is, all the data's noisy, so we were reluctant to publish," McDaniel told me from his office at Sandia. Besides, he added, "publish or perish is not a problem for government employees."

The "noisy data" that McDaniel referred to, however, is what some scientists just call bad data, i.e., data points that are literally scattered all across the graph. Whether the scattering was because of false signals from equipment, background radiation, or some other cause was not quite clear, but not all of McDaniel's tests were yielding the same results. And without access to final data or details of the experimental setup, no one could challenge those results. All they had were his unsubstantiated claims.

McDaniel carried out a total of six experiments during 2003, including three at the Air Force's facility in New Mexico, which he described as something like "a big, continuously running dental X-ray machine." The first experiment at the Air Force lab went well, according McDaniel, and showed 15 percent triggering—much higher than the 4 percent reported from the original dental X-ray.

But that was just one experiment, and Martin Stickley, in touting McDaniel's results, failed to mention what happened in the others. In one case, the "continuously running dental X-ray machine" had a meltdown. First the power supply went out, and then the electron tube broke as well. McDaniel's team had to move the target. Perhaps they moved it behind the shield, McDaniel suggested, because after that, the experiment didn't register any triggering. A third experiment at the lab also didn't register any triggering and succeeded only in melting the plastic-encased hafnium. McDaniel had to send the hafnium back to Hill Roberts in Huntsville, Alabama, and order a new sample.

Then there were three experimental runs at Louisiana State University. The first experiment, according to McDaniel, "seemed to corroborate Carl's results very well." The second experiment, however, did not. The third experiment had yet other difficulties. Stickley dispatched independent observers to oversee the experiment, and in a partial repeat of Collins's experience with Nick Stone, a conflict ensued. "They raised so many spurious objections that I still haven't run them all down," McDaniel explained bitterly.

McDaniel later reported that experiment a success.

In other words, of six experiments conducted that year, only half yielded positive results, and none of those results— good or bad—were published. In another troubling trend, independent observers dispatched to watch the experiments were only finding flaws.

But McDaniel was unperturbed. Remembering the good experiments over the bad, McDaniel said that Collins's detractors were simply nursing bruised egos because they were unable to repeat the original results. "The point I want to make is Carl's experiments have led where to look in time, energy, and flux level," McDaniel told me. "And a lot of the sniping is that he doesn't have a very good data because he had to take data over wide regions."

The problem was, it didn't seem like anyone had any good data.

Collins, McDaniel said, was simply a superb scientist. "I don't want to make this into an 'I love Carl' thing," he said, but the problem is people just don't understand him. Collins's writing could be confusing and people often have to read his e-mails three times over to understand what he's talking about, McDaniel explained.

Indeed, Collins's e-mails did have an odd tendency to rely on the passive case, sometimes treating nouns and verbs as optional additions to extended sentences. ["No experiment reported as positive has ever failed to be confirmed and no critic has ever tried to do THE SAME EXPERIMENT," Collins wrote me in a typical e-mail.] The end result, I had found, was often quite baffling. But McDaniel's defense of Collins's writing compared starkly to what other scientists described as merely unintelligible sentences.

When I asked McDaniel about the potential of isomers to be a superbomb, he at first denied being focused on weapons. Strangely, McDaniel, who had briefed the isomer weapon to the JASONs and had pushed the isomer weapon to DARPA, the Air Force, and Sandia, suddenly wanted to talk about engines. He thought hafnium would make for great aircraft engines, he said. Slightly radioactive, he conceded, but efficient.

But DARPA had never once mentioned propulsion, and rather, had explicitly stated its interest in a weapon, I pointed out to McDaniel.

Finally, he conceded that, yes, an isomer weapon was a possible application, but that was ultimately a political question. The United States went through the same argument in the early 1950s with whether to build the hydrogen bomb, McDaniel said. "People are fighting the isomer issue out in the press because they're worried about the political implications of another superbomb," he argued. But in essence, it's like the medieval debate over outlawing the crossbow.

"The issue is, in a free society, we need to know what's possible," he said. "We don't need to use it in a war, but we need to know if it's possible."

McDaniel's assertion about weapons and a free society was not unlike that of Collins, except that Collins would barely acknowledge the Pentagon's interests in his work as a bomb. In a phone conversation with Collins, he did, in an offhanded manner, recall to me that a senior Pentagon official had visited his lab. A man named Hopps, he said. John Hopps, an official in the Pentagon's defense research and engineering division, paid a visit in the spring of 2003—shortly after the DARPA isomer project got started. Hopps, a deputy to the Pentagon's chief technology officer, was concerned about the critical reports, Collins said, and "wanted to see first hand what we were like."

I STARED AT the phrasealator and the phrasealator seemed to stare back at me and I knew the interview was going nowhere fast. The office windows were tinted yellow—a special protective coating meant to block possible surveillance—and it cast a sickly parlor over the room. I had waited over a month for a thirty-minute interview with Ronald Sega, the Pentagon's chief technologist, and I only wanted to ask one question: Why had he sent his deputy, John Hopps, down to Texas to visit Carl Collins's gamma-ray lab? But sitting in his well-furnished Pentagon office, first I had to listen to Sega, whose official position was as the director of defense research and engineering, talk about DARPA's phrasealator.

It was show-and-tell time in Sega's office and an assortment of Pentagon toys were on his desk. Would I like to see a video showing a test detonation of the thermobaric bomb on his wide-screen TV? No, I replied, I'd seen that clip before, many, many times. The three-minute video—often set to music—featured the Pentagon's tunnel-blasting weapon smashing a bunker and was a mainstay of recent military conferences. No interest?

It was on to the next item in the showcase, DARPA's phrasealator.

Ronald Sega, the Pentagon's top military scientist, and nominal overseer of DARPA, smiled as he pointed at the phrasealator. Despite its ominously silly name, the phrasealator was not actually a weapon, but a universal translator for rogue languages. It looked a lot like a bigger, clumsier version of the universal translators selling for ninety-nine dollars in airline catalogues. It translated Urdu, Arabic, and Pashtu, spitting out such friendly phrases as, "How many minutes ago did it happen?" in clear, if somewhat mechanical diction. Sort of a Speak-and-Spell for Special Forces.

Sega wanted to talk about great things the Pentagon had produced for the war on terrorism, and the phrasealator was indeed an example of a DARPA invention that was hurried to the field post-9/11. While no one knew how much it was being used in Iraq and Afghanistan, Beltway officials lugged them out at every public opportunity they got. Sega wanted to talk about his phrasealator, I wanted to talk about isomer bombs and it was becoming clear who was going to win the battle. Sega was a former astronaut who had spent a few days on the ramshackle Russian space station Mir about a decade back. Like every former astronaut I'd met in Washington, he was a handsome man with a square jaw, nice smile, and a good suit. And perfect hair. He spoke in pleasant platitudes and always looked as if he were posing for the next publicity shot. It was hard to look at him without imagining him in a flight suit with all the movie characters from the *The Right Stuff,* or perhaps *Space Cowboys,* surrounding him.

If DARPA was supposed to be full of crazy and zany scientists, Sega's office in the E-Ring of the Pentagon was the adult supervision, and the grown-ups had heard that DARPA was playing around with isomers. Moving the conversation away from the phrasealator, I asked Sega about his thoughts on isomers and Hopps's visit to Texas. Sega frowned a bit, resting one plaintive hand back on the phrasealator.

"I haven't looked at it in any detail," he replied with a bit of a sigh. "Dr. Hopps has a background in physics, and he's taking a look at it." Sega paused, and then continued. "And I, uh, don't know, uh. . . ." He paused again, laughed nervously, glanced down at this hands, and then said, "I don't know where he comes down on it."

But what did Sega think about the DARPA project?

"The approach, as I understand it, is to look at a test that would answer questions," he replied. "That's part of the process."

"The process?" I asked.

"The process of physics that is going on, and I, uh, so I would . . ." Sega continued, his voice trailing off at the end. "So, our expert here, who is an expert in that particular area of physics, will do the details on that."

"Did he give you some feedback yet?" I asked, wondering what Hopps would have learned from a brief visit to Collins's dusty little lab in Texas.

The former astronaut ran a nicely manicured hand over the phrasealator nervously. "No, not yet. No, no, he's off looking at that," he said, somewhat disjointedly adding the words "good background." Presumably he was speaking about Hopps.

Sega sat silently, staring wistfully at the phrasealator placed on the conference table between us, and I realized the conversation about isomers would have to come to an end. Next, he wanted to talk about DARPA's water pen, a magic-marker size water purification system for soldiers in "tactical situations," the latest euphemism for a war zone.

BUT IN REALITY, more was going on behind the scenes than Sega was letting on, or perhaps knew, and it wasn't just about isomers, but the entire DARPA portfolio. DARPA had long resisted oversight, even when things were getting out of hand, and isomers were leaving a lot of officials mystified.

Concerns about DARPA's investments had been brewing for some time, but the public brouhaha over its Information Awareness Office—with its terrorism stock market and alleged plans for domestic surveillance—was the final straw. In October 2003, the Pentagon's chief acquisition official, Michael Wynne, issued a directive ordering an outside advisory panel to review DARPA's entire investment plan. The purpose of the review, Wynne wrote, "is to confirm that DARPA has advanced research projects based on sound, proven scientific and technological foundations, practices and methods" and would actually develop things useful for the Pentagon.

But for many involved in the hafnium program, things were about to take another unusual turn, because even if the scientific community considered the isomer bomb a joke, a number of people in high places in the world of national security were taking it seriously. For starters, there was Paul Robinson, the head of the Sandia National Laboratories. Robinson, while no longer working as a research scientist, was certainly, from a political standpoint, one of the country's most influential nuclear weapons advisers.

When I spoke with Robinson over the phone in late 2003, he told me that as a former experimentalist, he liked the romantic notion of proving the theorists wrong. Robinson also told me that he was fairly confident, based on Patrick McDaniel's research, that hafnium triggering was real.

"I suspect you'll keep looking at the triggering until you get it firmly established that you can do it," he had said.

Of course, by that point, he added, hafnium research "would probably be made classified, and you wouldn't read about it." Robinson's remarks proved prescient, because even if Pentagon officials were scratching their heads over DARPA's isomer obsession, bureaucrats in the State Department's Nonproliferation Bureau were tracking with interest a series of European and American news articles about the coming isomer arms race. Why, they wondered, was everyone talking about isomer weapons out in the open? In the fall of 2003, the State Department summoned Martin Stickley, DARPA's program manager, to talk about the isomer weapon. Asked if the triggering effect was real, Stickley apparently told the State Department officials that not only was it real, but that he personally had witnessed triggering at a recent experiment in Europe.

The State Department officials, perhaps not quite understanding that observing a hafnium triggering experiment was like watching grass grow, must have thought Stickley's eyewitness testimonial was of a massive explosion. Following the meeting, Phil Dolliff, a State Department official, sent out a slew of e-mails to government scientists marked "official use only" and the subject heading: "The Nonproliferation Bureau at State requests your assistance in assessing whether Isomer Research is sufficiently classified."

Scientists who had been involved in the work at Argonne were apoplectic. The idea that the government would make a nonexistent weapon classified struck them as silly, and also as dangerous if it meant that the scientific debate was going to become shrouded in secrecy. John Schiffer, one of the Argonne scientists, was in a bit of shock when he got the news, sending me an urgent e-mail expressing genuine outrage over the latest development. Classifying isomer research would only further remove hafnium from scientific debate, he told me.

Donald Gemmell, Schiffer's colleague at Argonne, did what he had done each time the isomer silliness had popped up next to concerns of a new arms race. He typed out a long, detailed e-mail attempting to explain why isomers were not the new nukes. In response to the assertion that triggering had been "observed" by Stickley and Collins in Europe, Gemmell wrote, "Where is the evidence upon which that assertion is based?" Pointing to the long line of publications and counter-experiments, Gemmell attempted to explain that Collins's various "triggering" results had been thoroughly discredited.

But Dolliff, who did not seem to have a scientific background, replied that his challenge as a "nonproliferator" was to weigh the risks to national security. "If there is even a small risk that this research could pan out into a new significant class of weapons then it is incumbent on us to be safe, not sorry and protect that information against terrorists and those states that support them," Dolliff replied in a tone reminiscent of language used to describe Iraq's weapons of mass destruction.

Gemmell was clearly hitting a brick wall. How could he make it clear that isomer triggering was no more than a fantasy? He wrote another note to Dolliff, this time trying to explain that, in fact, knowledge about isomers was well advanced. Both the physics of isomers and the expected response to X-rays was a well-understood phenomenon, Gemmell wrote. This was not two scientific groups squabbling over some unknown, but a matter of scientific fact.

The curious back and fourth between bureaucrat and scientist continued for a number of days. Still unable to grasp the essence of the problem, Dolliff finally asked Gemmell whether isomers could in fact store enough energy to create the near nuclear explosions that had been cited in various press articles.

Gemmell responded with a bit of frustration, attempting to explain that all materials store energy, and the question is always over whether that energy can be released quickly, and whether it can trigger a chain reaction needed for a bomb. In theory, hafnium-178 is a high-density material, meaning it stores a lot of energy in a small amount of material, in fact, over fifty-thousand times the energy in the equivalent amount of TNT explosive, Gemmell explained patiently. But there are lots of so-called high-density materials, he pointed out, including sugar.

Sugar, Gemmell suggested, has an energy density four times that of TNT. The problem is that sugar doesn't "trigger," or release energy, easily, Gemmell explained. "178Hf does not release its energy easily either. It takes several decades. So sugar and 178Hf are not explosive materials."

Neither was Gemmell the first one to make this basic point. A jelly donut, as the JASONS had pointed out back in 1999, has about the same energy density as TNT.

"Classifying the work at this stage would serve to protect this waste from public scrutiny," Gemmell wrote at the end of his e-mail. "I would sooner see us look for ways to trigger a chain reaction in sugar. The material is readily available, is not radioactive, has an energy density greater than TNT, and is about as likely to work as 178Hf!"

Gemmell never received a reply.

It was late fall of 2003 and the final meeting of the hafnium production panel, which was supposed to determine if enough hafnium could be produced to actually build a bomb, was only a few months away. And as weapons inspectors combed the Iraq countryside in a hapless pursuit of weapons of mass destruction, the U.S. government's proliferation experts in Washington were growing worried about rogue hafnium bombs.

Isomers, it seemed, had finally hit prime time.

11

THE MOTHER OF ALL DIRTY BOMBS

THE TRIP TO Carl Collins's lab in Texas didn't start under the most auspicious circumstances. Collins and I first discussed the possibility of visiting one of his hafnium-triggering experiments. That, however, proved unusually problematic. Collins was working exclusively at foreign synchrotron facilities, and that posed certain difficulties.

The Japanese, it turned out, were extremely nervous about the press, Collins said, and the Swiss, on the other hand, had simply decided they didn't want to be involved in isomer weapons. That left several X-ray facilities in the United States—including the one at Argonne, as well as others in California and in Louisiana. But for whatever reason, Collins preferred to do his research abroad. With the Japanese experiments out of reach, that left the option of visiting him at his laboratory in Texas. At least I could the see the site of the original experiment, I thought, or perhaps get a peek at the now-famous dental X-ray machine.

No sooner was that issue settled, however, then Collins immediately got into an argument with the university's public affairs office, which wanted to make sure that any visiting reporters also wrote about the university. Everyone, it seemed, wanted some publicity; unlike foreign laboratories, the University of Texas at Dallas had no problems being associated with weapons. For Collins, however, this seemed to be stealing attention from his work and prompted a flurry of increasingly angry e-mails directed at the university's public affairs office.

By the time I flew down to Dallas in January 2004, a compromise was struck. The public affairs office would get an hour or two of my time at the beginning so I could learn about the marvels of the University of Texas at Dallas, and then I would spend the rest of my visit with Collins. So, the first two hours of my visit to Collins's lab consisted of impromptu lectures by the dean of graduate studies—essentially Collins's boss—and the campus director of public affairs. And this is what I learned about the university: Conceived initially in 1961 as a graduate research center by scientists from Texas Instruments, the university was a magnet for science majors and high-caliber students who wanted to study close to home. The four founders—Eugene McDermott, Cecil Green, J. Erik Johnson, and H.B. Peacock—were concerned that Texas Instruments had to go outside the Lone Star State to recruit engineers. They wanted a university that could train local Texas talent in engineering and science. After all, who needs Asians and Indians if native Texans could do just as well? The center was conferred official status as a University of Texas campus in 1969; in return the lands and building were donated to the state, and the University of Texas at Dallas was born.

With little landscaping other than the occasional oasis of trees potted in concrete, the campus—a collection of low-slung, prefab-looking buildings—blended seamlessly with the tract housing and strip malls of suburban Dallas. Collins's Center for Quantum Electronics, a ramshackle one-story building, sat at the far corner of the campus. The building was so far removed from the rest of the university that it seemed almost as if Collins had simply set up shop there one day without anyone noticing, and his building was eventually absorbed into the campus the way expanding suburbs eventually meld into cities. The offices in his building were filled with random bits of furniture including a torn couch placed for no apparent reason in the bathroom and an ancient television set up on a leaning coffee table in the main conference room. Even Collins joked a bit about the state of his lab. University maintenance seemed to have forgotten about the center, and in the summer, he had to wear a cowboy hat to shade his head from the torrent of frigid air that blew out from the broken ventilation system.

The university had a physics department that was separate from Collins's lab, but there was no evidence that Collins had any contact with them, or vice versa; and in all the published papers I reviewed that Collins had written, no one else from the university's physics department ever coauthored anything with Collins. His colleagues all appeared to be from abroad. It was as if Collins had created his own little alternate physics department within the university, and no one had ever bothered to notice.

One of Collins's closest colleagues was a French scientist named Jean-Michel Pouvesle. After I had spoken with Collins the first time, he suggested I get in touch with Pouvesle, who struck me as a perfectly reasonable man when I first contacted him at his office in France. He was devoted to basic science, he told me, and not at all interested in bombs. It pained him to no end that hafnium triggering had "devolved to name calling" and was showing up on skeptics' Web sites, right next to reports of Yeti and Loch Ness Monster sightings.

"Should you fight scientific evidence by saying it's bad politics?" he asked. "Should you take questionable science, and say it could be a big bomb? Maybe that's bad politics and bad science."

Fair enough. Pouvesle, in fact, spoke quite convincingly about hafnium research.

"This is a very beginning," he explained. "There may be applications in medicine, or in transport of energy. For example, we may imagine this will enable us to send energy to a remote place on Mars. Or it could be used for a space-craft, because if you can store energy and release it little by little, with a small device, you can send this energy far from earth. It could be used for medical applications—for treatment of cancer, or even you can imagine uses for sterilization. I think there is a big struggle in United States around nuclear bombs. . . . People can imagine what they want. We are far from that."

Well, I asked, how far are we from bombs, or any other applications?

"The future is not tomorrow, it is in tens of years," Pouvesle told me.

It was that sort of sober, scientific evaluation that accompanied me when I arrived in Texas, thinking that perhaps, even if hafnium triggering was wrong, I should at least give Collins a chance. Maybe all of this was just an esoteric scientific controversy over an unexpected experimental result. After all, wasn't open-mindedness the basic idea behind scientific discourse?

When I arrived at his office, Collins jumped up to greet me like I was an old friend returning from an extended trip. And after I was done learning about the university, he offered up one of his graduate students, a bored-looking Romanian named Catalin Zoita, to guide me around the offices and lab. Zoita, who didn't talk much, proved a challenging guide, and I couldn't figure out exactly what he did, other than alternate between extended periods staring at a computer screen, punctuated each half-hour by cigarette breaks outside the office. If Zoita thought he was working on groundbreaking physics, or was at the center of a new weapons project, he certainly didn't show much enthusiasm.

Zoita was one of Collins's two graduate students when I visited. The other, a robust-looking Norwegian woman named Elin Aronsen, also didn't have much to say, and spent most of the time hiding out in a back office where Collins stored his old technical reports, which were gathering dust on metal bookshelves. The only other occupants of the building were Farzin Davanloo, a preternaturally shy Iranian scientist who worked on pulsed lasers, and a young Romanian woman named Delia, who was working as a secretary but admitted she wasn't even being paid. Pouvesle, Collins's French colleague, was visiting the lab during my stay. The only other person to show up at the lab in the two days I spent there was a confused young student, who entered the building by accident. She was looking for her economics course, she explained, glancing around nervously, as if she had just stumbled into the wrong house.

On my first afternoon at the lab, Zoita guided me around the main office building, explaining in a Peter Lorre accent what the various laboratory artifacts signified. For example, he pointed to two upturned Styrofoam cups behind a glass case. The first cup was marked "Dr. C's memorial target holder."

Zoita explained that this was the very Styrofoam cup that held the isomer target in the 1998 experiment. Next to it was a second, identical cup that read, "A cheap imitation." It was presumably a joke, but Zoita didn't laugh, or even look amused.

We walked behind the main building and into the lab, a boxy aluminum structure, only half of which belonged to Collins. The lab was choked with dust, and I later learned that Collins refused to allow university custodians to even enter the building, which housed two decades' worth of experimental equipment. Surplus from the nuclear weapons laboratories, copper wire, random screws, and car parts bought on the cheap all crowded the little lab. Even the metal sign out front, sporting the Greek symbol for gamma, was molded from a discarded highway marker. The layers of dust, broken knobs, bits of wire and discarded electronics were all just relics. Almost none of the equipment was used anymore.

We arrived at the main section of the lab, and on the ground next to the site of the 1998 experiment was the discarded head of a dental X-ray machine, still attached to its swivel arm. Lying on the dusty floor of the cavernous lab, amid flickering computer lights and industrial machinery, the disembodied X-ray arm appeared ready to spring back to life like a revived limb of the Terminator.

"Is that it?" I asked. "Is that the dental X-ray?"

Zoita shrugged his shoulders, and went out to smoke a cigarette, leaving me to stare at the X-ray arm.

It wasn't *the* dental X-ray device, Collins later said, when I asked him about it. It was just one of several dental X-ray machines he had bought over the years. He said he wasn't even sure where the original one was—perhaps somewhere amid the dust and discarded equipment.

Other remnants of the original experiment were still there, however.

The audio amplifier used in the original experiment—the same kind you would find at a Motörhead concert—lay encased in cinderblocks below the test bed, at rest in its final burying place.

Back in the office, a faded homemade poster hung in the front entrance. It looked a bit like one of those science posters students are forced to make in high school toward the end of the year, when the teacher inevitably runs out of things to assign and wants to keep everyone busy. In washed-out brown marker, large block letters spelled out "Motivation." And below that: "The greatest densities of energy which can be released without a nuclear weapon are stored in certain isotopes as metastable states called isomers."

At the bottom, the text read: "Now underway are preparations for acquiring and testing a sample."

According to the poster, Collins and his colleagues had performed thousands of experiments over the span of a decade to find the best isomer candidate. Collins shook his head and smiled ruefully as he explained why the material was only identified as isotope X.

"It was politics even then," he said.

Collins had been looking at many different isotopes, and each time he found a promising one, his critics called it impossible, and would try to undermine his work, he claimed. "After that, we decided we're not going to tip our hand until we just try it."

That secrecy extended even to hafnium, which was nowhere to be seen in Collins's lab, and he was reticent on the entire subject. When I asked him where it was, he avoided the question, and when I enquired how he was able to send it to faraway places like Japan, he replied, "I'll tell you how I do it if Jeff Carroll tells you how he sends his."

Collins's claims of persecution and need for secrecy had a tendency to work well with non-scientists, who weren't able to followed the scientific arguments. So, too, did his convoluted explanation of experimental results. After all, who other than physicists could follow some of the more obscure arguments over "K-mixing levels" and "vacancies in L-shell electrons"?

Harder for Collins, on the other hand, was answering some of the common sense suggestions put forward by other scientists. Collins, for example, acknowledged that some scientists have asked whether he considered doing what sounded like a relatively simple null experiment with hafnium: irradiating a sample of ground-state hafnium and comparing that with the results of the hafnium isomer. It would be, in a certain sense, like comparing the effectiveness of aspirin to sugar pills. If for any reason the "triggering" lines that Collins was seeing were an experimental artifact, using regular hafnium would be one way to demonstrate this fact.

Not surprisingly, Collins described that suggestion as just "a lot of trouble." Why bother do to a null experiment,

Collins asked? "First, there's nothing to see, so you're not going to produce anything," he said.

Then there was the problem of using those fancy X-ray facilities. The managers who controlled the facilities would question the point of doing a null experiment, he claimed. "How are you going to present that as a worthwhile use of the time they gave you?" he argued.

Anyhow, Collins said he discussed the idea with Martin Stickley, the DARPA sponsor, and Patrick McDaniel, the Sandia scientist, and since both of them believed hafnium triggered, neither of them thought a null experiment was necessary. Why question your own results if you're convinced you're right?

But that brought Collins back to the touchy issue of the Argonne group, which had repeated his experiment, and had not found any evidence that hafnium triggered. The scientists involved in that collaboration were often referred to as "world class," "the nation's top experimentalists," or simply "superb scientists." Collins, by contrast, described them as "failures, with a capital F."

With evident glee, he continued, "They said in their own article, 'we failed.' The failure group has always failed."

I glanced around the table. Collins's remarks were not exactly what one would expect out of a scientist, even one embroiled in a disagreement with others in the field. But none of his colleagues uttered a word. Zoita, the Romanian graduate student, fidgeted with a cigarette, and Pouvesle, his French colleague, shuffled his feet and stared down at the table. Collins was beginning to sound a bit manic.

"We might suspect that they don't know how to do it," Collins continued. "Maybe there's a defect at the Argonne synchrotron. It's quite old. Since nothing ever succeeded at Argonne in this line of work, there's no reason why it should succeed."

Pouvesle, Collins's French colleague, sat mute as Collins continued to rail against his enemies. Nor did Pouvesle, who had told me that isomer applications were years away, even blink an eye when I asked Collins about DARPA, which wanted a bomb in just a few short years. Collins said he didn't see anything wrong with the Pentagon's plans. It wasn't his problem.

At dinner that night at a New Orleans–style Cajun restaurant in a Richardson strip mall, Collins continued with his elaborate stories of foreign travels, lurking enemies, and triumphs over adversity. His colleagues sat enraptured, with all attention focused on Collins as he regaled his audience with stories of trips to India, Saudi Arabia, and Japan, and tales from his days living in Romania. Doina, his cherub-faced wife, also delighted in his stories, occasionally interrupting to provide details and color. He talked about being fired from teaching undergraduate physics; a result, he claimed, from challenging his students to think critically about nuclear power. And he railed against the scientists at Livermore and Argonne who were trying deny him his place in history for discovering hafnium triggering.

The world was full of people who didn't understand him, enemies who were out to get him, but also people who adored him, like his wife, Doina, who piped in, "It's always been like this, but sooner or later they will recognize him and his results."

The table nodded in agreement.

Collins laughed, and then ended dinner with another story of persecution, recalling how he was invited some years back to give a lecture on isomers at Sandia. He was "confronted," he said, by dozens of scientists who supported Mort Weiss, the physicist from Livermore who criticized his work. Weiss, according to Collins, got up during the lecture to challenge his claims that he could trigger isomers with a simple X-ray machine.

Collins's reply, as he recalled: "Well, Mort, that's because you don't have any isomer."

Doina, Collins's wife smiled and laughed. After all, she owned the isomer sample.

THE NEXT DAY, back at his lab, Collins was in top form, at least until we got to the topic of weapons. Collins's claim was still that his only interest in isomers was to build a gamma-ray laser, a device that would have commercial and civil applications. That was a curious assertion, however, for a scientist whose almost exclusive source of funding for the past two decades had been the Pentagon. I looked around again at his colleagues, who were of course silent. In over sixteen hours of interviews at the lab, they had uttered barely a word, except to agree with some statement or other that Collins had made.

"Actually, I don't know anything about what the Pentagon is doing," Collins said at one point when I pressed him on the issue.

Collins had two distinct sets of briefing slides: those on his Web site and that he had showed me on my first day at the lab, which depicted just his experimental results, and a second set, which he used in military briefings. The second set talked about bombs.

He didn't know I had a copy of the slides that he had presented to DARPA at the May 2003 meeting of the hafnium production panel. His presentation from that day showed a picture of a man hitting a golf ball across a field, with a mushroom cloud rising in the distance at the end of the ball's arc.

The caption read, "A golf ball filled with the isomer would have the energy of 10 tons of explosive."

Collins was fond of showing that picture in Pentagon briefings, according to those who knew him. But he was visibly uncomfortable when I asked about the diagram, which was briefed at a closed meeting. He said apologetically that the "sponsors," DARPA, has asked him to make an illustration to show hafnium's potential. Collins was an avid golf player.

It was the final hour of interviews at the lab, so I asked Collins one last time how he felt about conducting an experiment whose results, if true, could lead to the next superbomb. The other members of his team grew silent and stared down at their coffee cups as Collins began to talk about the need for science in a free society, about the medical applications, and then paused.

"I guess I don't feel anything. As some point, I'll retire and go play golf," he said with a smile.

AT CARL COLLINS'S office in Texas, a clutter of paper shrouded his desk. A portrait of a beautiful young woman—an old photo of Collins's wife, Doina—stood out amid the chaos. Nearby was a copy of a novel by Dallas author Payne Harrison, who had written a number of techno-thrillers. The image of his desk and the book stuck in my mind long after I returned to Washington. Collins didn't strike me as the type to read potboilers, so when I got home, I logged onto Amazon.com.

It looked like Harrison hadn't published in many years, and his older paperbacks were selling for about a dollar apiece. One was going for a mere cent, plus shipping costs. A real bargain, I thought, and ordered it.

When the book *Thunder of Erebus* arrived in the mail, I initially tossed the weather-beaten paperback on the coffee table, returning to it only later that evening. As I began to read the stilted prose characteristic of military thrillers, I suddenly realized I was reading a Cold War potboiler about the isomer bomb: a fictional isomer called rubidium-86, gamma-ray lasers for Star Wars, and then DARPA's development of a new, conventional superweapon based on isomer triggering after Star Wars is canceled.

In the book, Russia decides that it must control the world's supply of rubidium-86 (which, in the novel, is in Antarctica) for fear of DARPA's secret isomer weapon. As if tearing a sheet from modern history, the Pentagon, fearing the Russians would build their own isomer weapon, decides to invade a country in the Persian Gulf.

"General, if the Soviets seize this rubidium deposit and fashion a scimitarlike graser weapon, and if our intelligence is true that the Soviets are quickly running out of oil, could they then seize the entire Persian Gulf and leave us powerless to break their control of our vital energy source?" says the national security advisor to General Rodger Whittenburg, the fictional head of the U.S. Armed Forces, on the dangers of an isomer weapon.

Wow, I thought. Harrison was a genius, or at least a possible candidate for CIA remote viewing. Here we were in 2004, and we really had invaded a Persian Gulf country, and we really were worried about Russian isomer bombs. Okay, there wasn't a clear link between the two, but it was still awfully close to reality.

Some years ago, Collins told me, Harrison hung around the gamma-ray lab for a few months "to absorb the culture." The author, a former tax accountant, shadowed Collins's team, eventually drifting out of the lab as quietly as he came in. The odd thing was, *Thunder of Erebus* was published in 1991—over a decade before DARPA funded work on an isomer bomb. Collins offered no explanation for how Harrison's book came so close to describing modern reality. A lucky guess? Or did Harrison, an astute observer, pick up rumors of a secret, classified military project to build an isomer weapon?

A good portion of DARPA's funding is classified, and the projects under the cloaked part of its budget are never revealed publicly. Among Collins's critics, there were rumors that his work was funded in the 1990s as part of an Air Force "black," or classified, project.

Classifying military programs, particularly those that relate to advanced science, tends to lead to two problems. The first is bad science. Without the ability to present scientific results in peer-reviewed literature and attend open conferences, the quality of science tends to degrade very quickly; without review, who is going to tell you a given idea is flawed? Science thrives on openness. That has been, and continues to be, one of the biggest problems with military science, dating back to the days of the Manhattan Project when J. Robert Oppenheimer was forced to battle the military bureaucracy to allow scientists to exchange as much information as possible.

The other consequence of classifying scientific research is somewhat less severe, but still problematic. It encourages conspiracy theories. The best example of that is the early work on radar-evading stealth technology. Billions of dollars were spent on the ominously bat-shaped B-2 Spirit bomber, which was shrouded in secrecy, allegedly so that the United States could maintain its edge over the Soviet Union. While the bomber, designed to be invisible to Soviet radar, was eventually rolled out to the public in 1988, the secrecy and disinformation surrounding the aircraft gave rise to conspiracy theories that the bomber was actually a cover for secret antigravity technology; a half-baked theory pieced together from disparate pieces of missing or incomplete information. In some versions of the story, the antigravity technology was obtained from a crashed UFO.

Was the B-2 part of a secret antigravity program? Absurd. Did DARPA sponsor a secret isomer weapon project in the 1990s? Unlikely. But that's the problem, when it comes to the "black world"—as classified programs are euphemistically referred to—there's simply no way to know. As for a secret isomer bomb, there's no proof—other than Harrison's prescient, if slightly silly, book—of its existence.

Fast-forward to the real isomer weapon in 2003. It was

only a few months after the start of DARPA's project, and
Martin Stickley was in a very bad way. Things were going
wrong, terribly wrong, and he knew it. Collins's experiments—despite his public claims—weren't reproducing
consistently, and Collins had refused to repeat the original
dental X-ray experiment. The independent observers Stickley
had hired were issuing only criticism, and he was under major
pressure from DARPA's director, Tony Tether, to produce
results. Tether had only agreed to fund hafnium if something
practical could be made out of it, and that meant Stickley
had to demonstrate triggering, and soon.

Collins was continuing to publish new experimental
papers, all of which claimed proof of hafnium triggering.
Yet each paper seemed to make new, and equally incredible
claims—not so much confirming his earlier experiments
as spawning entirely new results. For example, in his early
papers, Collins claimed the primary proof of isomer triggering was enhancement of gamma emissions at 495 KeV.
In later papers, that claim switched to the 130-KeV line, and
then later to yet other claims. In his first paper, he claimed
X-rays with energy levels around forty thousand electron
volts, or 40 KeV, triggered hafnium, later, that changed to ten
thousand electron volts, or 10 KeV. And so it went.

By 2002, Collins also was claiming in European journals
that hafnium triggering was the result of an exotic nuclear
process called nuclear excitation by electronic transition, or
NEET for short. In NEET, as one scientist explained to me,
the electrons surrounding the nucleus of the atom are of a
much larger size than a nucleus, perhaps creating a way to
transfer atomic energy into nuclear energy.

"What is hoped for is that some sort of accidental, or
'magical' condition where the atomic processes resonate with
nuclear excitations," Jerry Wilhelmy, the Los Alamos scientist, explained to me. NEET was, of course, a real process,
but the problem was, as scientists were quick to point out,

is still couldn't explain the levels of triggering that Collins claimed. One Russian scientist, Eugene Tkalya, followed up with a whole article in *Physical Review* ridiculing the idea. The nuclear and atomic levels are quite narrow and the probability of striking this magical resonance was too low. The Russian scientist pointed out that even if Collins was right, his experimental results were still three orders of magnitude larger than what the NEET process would explain.

In short, Collins's results still violated known physics.

But none of this scientific quibbling mattered to Stickley, who had promised to give DARPA's director a bomb. Over dinner one evening with Ehsan Khan, the Department of Energy official working at DARPA on hafnium, Stickley apparently conceded a new problem. Tether had briefed the isomer bomb to Secretary of Defense Donald Rumsfeld, who thought the idea of an isomer bomb was great.

Rumsfeld wanted a prototype delivered within about eighteen months.

At that point, the world's supply of hafnium consisted of a few micrograms, and the Pentagon's chief wanted a bomb and he wanted it soon. Khan called James Carroll at Youngstown University, and made it blunt: someone had to demonstrate triggering of something soon, or the whole program was dead.

Carroll tried to convince Khan that maybe low-energy triggering of hafnium wasn't the way to go. There were other isomers they could look at perhaps, or maybe they could try to demonstrate triggering at higher energy levels. Of course, the other possibilities probably wouldn't make for a great bomb.

Khan promised to think about it, but nothing seemed to come out of it. Maybe it was too late to go back. Maybe Khan just believed so strongly in the hafnium bomb that criticism or science didn't even matter. Whatever the case, the second meeting of the hafnium production panel was fast approaching. It was a critical juncture for the hafnium bomb. The January 2004 review would determine whether the hafnium isomer could be produced at an affordable level.

What did Tony Tether, the head of DARPA think of all this? Surely a man who controlled billions of dollars in federal science and technology spending would approach such a controversial subject with a touch of skepticism. But Tether, it turns out, was pretty excited about the hafnium bomb, which in a November 2003 interview with Amy Butler, my colleague from *Aviation Week*, he described as still being in the "exploratory phase." DARPA's interest, he said at the time, was part of a long-standing Pentagon effort to make big bombs into mini-bombs.

"We spend a lot of money on taking explosives and trying to make it two times or four times more powerful, because you get a corresponding decrease in size," he said. "If hafnium were able to hold up to its promise, we could get on the order of 100 to 1,000 times more explosive out of the same volume, which means we could take a big bomb and make it into a very small bomb."

DARPA's goal, Tether continued, was to confirm that hafnium could be triggered, and then decide if enough could be produced affordably to use in a bomb.

And what about the scientific controversy, Amy asked him?

Tether merely shrugged his shoulders.

"Hafnium has been something that has been controversial over the years," he suggested. "You could take 100 physicists and probably they would line up 50 on each side."

Isomer triggering, Tether said, was a lot like the discussion that took place in the 1930s about nuclear fission—a concept that many in the military establishment had also once dismissed as science fiction. The idea of an atomic bomb, when first discussed during World War II, was also quite controversial. Hafnium wasn't a nuclear weapon, he quickly added, but it was a similar sort of controversy.

What would the Pentagon do with such a weapon?

Hafnium bombs would be extremely useful in many situations, Tether replied. They could be packed into unmanned aircraft, for example. And then there was the increasingly urgent issue of deeply buried bunkers that housed weapons of mass destruction or the heads of rogue states.

"Short of using a nuclear weapon, this hafnium bomb would be really good," Tether declared.

I HAD BARELY introduced myself to William Herrmannsfeldt over the phone when the soft-spoken Stanford physicist deftly steered the conversation to cold fusion.

"Cold fusion, when it happened, affected my very inner psyche," he said, when I reached him at his office in Menlo Park, California. "I couldn't have a civil discussion with people on it, I had to break off the discussion and go away."

At the height of the cold fusion controversy, Herrmannsfeldt explained, it was hard to have a rational dialogue. He very quickly ceased being cordial with the proponents of cold fusion.

At one point, Herrmannsfeldt got a call from an investor, asking for his advice on whether to invest in cold fusion. The investor offered the ultimate gambler's fallacy: even if cold fusion had one in a million chance of being right, wasn't it worth the investment? No, Herrmannsfeldt replied, because the chances were even much worse than that.

For Herrmannsfeldt, isomer triggering was the same thing. DARPA often pursues some eccentric research ideas, but they're "not supposed to be doing perpetual motion machines and antigravity machines," he said.

There's a difference, Herrmannsfeldt argued, between far-out ideas and just plain stupid ones.

And the hafnium bomb was worse than cold fusion in

one way. At least cold fusion, if true, would have produced something useful, like a source of energy. Hafnium, on the other hand, was being promoted as a superbomb. Worse yet, hafnium's radioactivity would make it a particularly dangerous bomb.

An isomer bomb, like any bomb, doesn't actually burn all its explosive content. So, an isomer bomb, for example, might have fifty thousand times the energy of a regular bomb—but that only accounts for about one-sixth of the material. In practice, for instance, a hafnium isomer bomb would leave behind one million times more residual radioactivity than a fission bomb.

"So most of it doesn't trigger and just gets scattered around. If this is in outer space, maybe you don't care, but over the city or golf course shown in the view graphs for the generals, it's not very nice," Herrmannsfeldt said.

"If you actually made a weapon with it, you'd have the mother of all dirty bombs."

But the broader issue was the science. Here were the nation's top scientists saying the stuff wouldn't work; yet the Pentagon was pushing forward and funding it. Worse yet, DARPA wasn't even willing to hear from the critics. Herrmannsfeldt explained that he was pushing for an independent review, but so far, no one in the Pentagon was listening. Asked about his role on the production panel, Herrmannsfeldt merely sighed. He hadn't been kicked off the panel yet, and what did he care if he was? He was a semi-retired physicist whose work for the DARPA program was unpaid. But he just couldn't believe that anyone would fund such a thing. And the production methods—what his work was focused on—were already showing that the material would be obscenely expensive to make, over a $1 billion a gram.

In other a words, a thirty-gram weapon of the sort that was being proposed would cost over $30 billion to make.

And that's not counting the costs to build the production facilities. The whole thing was ludicrous, and no one would listen. And rather than simply disappear, as Herrmannsfeldt hoped, the hafnium bomb was taking on a life of its own. In November 2004, *Wired*, the Silicon Valley magazine of technology and culture, ran a sort of "what's in/what's out" column, declaring that good old-fashioned uranium was "tired" and plutonium was "expired." Hafnium, on the other hand, was "wired."

Herrmannsfeldt was coming to the Washington area for a second meeting of the hafnium production panel, and I suggested we meet up in person. He was staying, it turned out, in the same Virginia hotel where I had met Collins just six months prior. When I arrived, Herrmannsfeldt was already standing in the lobby, ready to go. To say that Herrmannsfeldt was the archetypical physicist would not quite capture the picture; but it was easy to see why Pentagon officials might prefer a down-home Texan to a West Coast physicist. With his cardigan, striped tie, and pocket protector, I could almost hear the collective groan of Pentagon generals at the prospect of being lectured by a bespectacled academic, ready to tell them their bomb wouldn't work.

It didn't matter to Herrmannsfeldt: he was a man on a mission. He gave me a quizzical look, and then launched into a tirade on the hafnium bomb. There was an unusual snowstorm in Washington that January and as we drove from his hotel to a nearby restaurant, the snow came down in thick, wet sheets. Sliding across the road in the middle of a snowstorm, past DARPA headquarters, Herrmannsfeldt was oblivious to the weather; he began to give me an abridged version of his life's story and how he ended up battling the hafnium bomb.

He continued talking even after we entered the crowded restaurant, where tired Irish music blared from the speakers.

Herrmannsfeldt pointed to my tape recorder on the table. "I guess I can say what I want, because that's never going to pick up anything," he remarked with a chuckle.

"Yes, it will," I said.

He shrugged his shoulders—after all, he was semi-retired. He took a modest sip of his beer, and began to talk.

Herrmannsfeldt was originally from the Midwest, where he earned his doctorate at the University of Illinois in nuclear physics. "After Illinois, I went to Los Alamos and we were there four years and I was in weapons research. Four years of that was enough. Nuclear weapons research is about as boring as anything can get in science. Nothing ever happens."

Of course, that wasn't entirely true. There may not have been many new breakthroughs in the 1950s, but there were certainly a lot of ideas—some of them strange—on ways to use nuclear weapons. One of the strangest was Project Orion, a spaceship that would travel to distant planets, an idea championed by nuclear weapons designer Theodore Taylor. While there have been many exotic proposals for space travel, Orion was truly unique. It was going to use nuclear bombs to power its flight.

Perhaps not surprisingly, Orion was one of the first projects to be sponsored in 1959 by a new Pentagon agency called DARPA.

While many scientists still remember Project Orion with a starry-eyed romanticism that conjures up visions of distant space travel, Herrmannsfeldt had less-than-fond memories of what sounded an awful lot like a crazy idea. At Los Alamos, where Herrmannsfeldt was working, one of the applications for the nuclear-bomb powered space travel, for example, was for satellites that would monitor Soviet nuclear tests. "I got involved in a couple programs detecting bomb tests in outer space," was the way he put it.

Herrmannsfeldt's memory of the lunacy, which began when he came to Los Alamos, was still quite vivid:

The notion was that the U.S. has started a moratorium on air [above-ground] tests just about the time I got there. Then, we started the underground tests. But we worried about what the Russians would do; there were some concerns that Russians would do tests in outer space and we wouldn't be able see them. I think it was George Gamow. Does that name ring any bells with you? Gamow was one of the great figures in physics and stellar evolution—how the stars form elements. Anyhow, it was Gamow and a couple of other guys, but Gamow came to Los Alamos when we were trying to plan this program, and he suggested the Society for the Preservation of the Virginity of Venus. We were going to make sure nobody tested any bombs on Venus.

So, Los Alamos was coming up with a scheme where you have the sun and you have the earth over here going around the sun. And you want to put six satellites in equally phased orbit around the sun so that if the Russians go out there and put something on the other side of the sun, you could detect it. These were supposed to be stationary—not stationary in the sense of stopping—but stationary in the sense of their position with the earth, like in orbit around the earth. It takes an awful lot of energy to do that. You don't just fire up a chemical rocket like you do today and because getting it over there and arcing at a finite time takes an awful lot of energy. So, one of the fathers of the hydrogen bomb, Stanislaw Ulam, invented Project Orion. That was going to take a rocket ship where you pop little atomic bombs out of the tail and go firing off into orbit. But this was a bomb testing facility and you're going to get there by shooting off bombs? At which point, I said, "Okay, I can work on that." I could never imagine it in a million years. There are a number of things like that in my lifetime. I could work on it, but I couldn't imagine it.

"Satellites around the sun?" I asked, a bit incredulous.

"Yes, but that's okay," he said. "The earth is a satellite around the sun, so at some level that's okay. But to get there—to do it to detect bombs—and to get there by shooting off bombs. It was too much of an oxymoron. I couldn't handle that," he said with a sigh.

So, in 1962, Herrmannsfeldt packed his bags, and together with his wife Marcia, he moved to take a position at the Stanford Linear Accelerator Center. It was the best decision he ever made, he told me.

No longer involved in research, Herrmannsfeldt's time was spent on what he freely acknowledged were political battles. Ironically, it was Tom Ward, a major contractor for DARPA's hafnium program, who called Herrmannsfeldt to invite him to be on the panel. With a PhD in nuclear physics, and a career that spanned four decades at the Stanford Linear Accelerator Center, a Department of Energy laboratory with three Nobel Prizes to its name and a long history of innovative scientific research, Herrmannsfeldt was a logical choice, particularly given his experience in accelerators, one of the possible production methods for hafnium.

It was probably the worst choice Ward ever made, Herrmannsfeldt later joked.

The problems started almost immediately after he contacted Martin Stickley and Ehsan Khan, the two officials in charge of hafnium.

"I begged Khan to invite the critics. Maybe I even threatened him, because this stuff was even worse than I thought."

Herrmannsfeldt spoke calmly and softly about his concerns. He jotted down equations to show how DARPA would never get any useful energy out of hafnium. Even if Collins's results were true, what could you do with them, asked Herrmannsfeldt? If you grant Collins's claims—that a 10 KeV photon could trigger 2.5 MeV in hafnium, there was still a significant problem because Collins, in a later paper, admitted that only about one out of every six hundred of the 10 KeV-volt photons being sent into hafnium actually triggered something. That meant the math was a lot worse, because now it would take 6 million electron volts, or 6 MeV, to release 2.5 million electron volts of energy, or 2.5 MeV. In other worse, well below breakeven.

But that even assumes hafnium triggers. And Herrmannsfeldt pointed out that Collins's "proof"—his famous 130 KeV line—was barely statistically present.

"If you look at his plot, you cannot look at his plot and conclude that he saw anything," he said. "But he looked at that plot and concluded that he saw something."

Herrmannsfeldt talked about the reviews and competing experiments. He acknowledged his political concerns about the program—he called hafnium "the mother of all dirty bombs" that would entice other countries to build nuclear weapons—but he based his argument on science. Herrmannsfeldt had been talking for over two hours, and paused, just for a moment, to reflect on the meeting he was scheduled to go to the next day. Maybe they'll just pack up, he suggested, and tell everyone to go home.

"I doubt that," I replied. I had never heard of a military program being spontaneously cancelled just because it wouldn't work. There was no precedent.

I dropped Herrmannsfeldt back off at his Beltway hotel, and asked him to get in touch with me following his meetings.

When Herrmannsfeldt called the next day, the news was not good. After the production panel meetings, he and Donald Gemmell, one of the Argonne scientists, had gone to visit the Department of Energy to complain about the hafnium bomb. They spoke to Gene Henry and Peter Rosen, both scientists who worked in the Department of Energy's physics division. The problem was, hafnium was a DARPA project so there wasn't a lot the Department of Energy could do about it, they were told.

The production panel itself had gone on almost the whole day, and then later, at dinner, Herrmannsfeldt had pulled Stickley and Khan aside to speak to them about his concerns.

"I complained about the lack of respect for scientific advice, major reviews such as the JASONs and IDA," he said after the meeting. "Martin [Stickley] then came back and not very politely told me DARPA was above such things, and 'could ignore any publicity' around the program."

As for costs, Stickley and Khan were advertising that the initial $1 billion a gram cost could be brought down to as little as $1 million, but Herrmannsfeldt was doubtful. Though the production panel had looked at several different methods for making hafnium, all of the options seemed absurdly expensive. For example, the capital costs alone to build the accelerators would run between $20 and $60 billion, not to mention the money it would take to actually produce the isomer.

Nonetheless, the end result of the meeting, according to Herrmannsfeldt, was that Khan and Stickley were enthusiastic that production costs could be brought down; Stickley was supposed to meet with Tony Tether, DARPA's director, to communicate precisely that conclusion.

The isomer bomb would go on.

DARPA'S FUNDING OF the hafnium bomb hadn't gone unnoticed in Washington, most importantly by the intimate band of nuclear weapons specialists who populate various government agencies. Some of the whispering centered around

why DARPA was funding anything related to nuclear weapons—far outside the agency's typical mandate. Nuclear and
nuclear-type weapons are usually left to the National Nuclear
Security Administration, an arm of the Energy Department
that manages the nuclear weapons complex, or to the Defense
Threat Reduction Agency, the part of the Defense Department that works on nuclear effects testing.

When I cornered Stephen Younger, a former Los Alamos
weapons designer and head of the Defense Threat Reduction
Agency, at a Washington conference, I wanted to know what
he thought of the DARPA program. I hadn't seen him since
the day I had first heard of hafnium back in the fall of 2003.
The Defense Threat Reduction Agency at one point also had
dabbled in the world of isomer triggering.

Younger had just given a speech in Washington supporting the development of new, low-yield nuclear weapons when
I tracked him down. Introducing myself briefly and explaining
my interest in isomers, he looked a little tired.

"The quick answer," he replied testily, "is that it's best for
me not to comment on a DARPA program."

He looked over at a nearby door leading out of the audi-
torium; a clear escape route. I took a side step, cutting off
his easy getaway.

"Okay, besides the DARPA program? What do you think
about the fundamental idea of a hafnium bomb?"

Younger glanced at his watch.

"It's a very hard problem," he said. "This hafnium isomer,
it's difficult to store and the energy is difficult to release."

And there's the safety issue, he remarked. How would
anyone get near a bomb that uses a material with a thirty-
one-year half-life. And if it did trigger, and it did chain,
why wouldn't it just all blow up at once? The idea of isomer
triggering is one of those parlor conversations that come up
every few years among weapons designers, he explained. It's
looked at, it's discarded, and then it comes up again.

But if it did work, would it be a nuclear weapon, I asked?

"No," he quickly replied. "The ends are substantially different. Nuclear weapons are in the megaton range."

So a two-kiloton hand grenade didn't count as a nuclear weapon.

Not everyone agreed with Younger's view. When I called up Peter Zimmerman, the former government scientist who had first tried to kill the hafnium bomb back in 1999, I got a very different perspective.

"It's a small nuclear weapon, but it's a nuclear weapon," he insisted.

Getting ready to move to an academic position in London, Zimmerman had given a lot of thought, of late, to his decades of work in arms control and science. On the phone late that Saturday evening, he began to talk about the movie *Dr. Strangelove*. The parallels between the "isomer gap" and *Dr. Strangelove*'s "mineshaft gap" were so easy as to fall into a worn cliché. But Zimmerman reminded me he grew up in an era when the movie was more than just a caricature of Cold War politics.

"Sometimes there's an advantage to having lived through these issues and see them close up," he said. "It gives you the opportunity to laugh at the really foolish stuff."

The isomer bomb is foolish, he said, but it's foolish in a dangerous sort of way if it pushes other countries to build real nuclear weapons in the hopes of deterring the United States from using a fanciful hafnium bomb.

"Most of the really strong supporters of hafnium don't have a background in physics," he said. "The whole thing is laughable. But I ask my colleagues, should we write a rejoinder and let this die a natural death?"

For a moment there was silence on the phone, as if Zimmerman was still contemplating the answer, before adding one last thought.

"I would let it die, but it keeps coming back to haunt me."

JOURNALISTS LOVE A good story about exotic new weapons. Perhaps they love it even more if it turns out to be true, but does it really make any difference in the end? In the 1990s, newspapers were awash with stories of "Red Mercury," a mystery substance allegedly being sold on the black market of the Third World, possibly by former Soviet scientists. Certainly in its claims of being a possible source for miniaturized nuclear weapons, Red Mercury was similar to hafnium. And for a while, Red Mercury became the nuclear weapons version of Elvis sightings. It was everywhere.

Depending on whose poorly sourced account you read, Red Mercury was highly enriched uranium, or some other mystery material that could be used to make pure fusion bombs. Some reports claimed the mystery compound was a derivative of Soviet nuclear weapons, or an additive to rocket fuel. Yet other reports claimed it was a by-product of nuclear reactors, and could somehow be used to trigger a briefcase-sized mini-nuke.

The International Atomic Energy Authority took a more critical view than the press, however. Or maybe it just tired of over a decade of Red Mercury sightings.

"Red Mercury doesn't exist," a spokesman for the Vienna-based organization told the U.K.'s *Guardian* in 2004. "The whole thing is a bunch of malarkey."

Perhaps the hype over Red Mercury explains why, in mid-2003, the Union of Concerned Scientists, a nonprofit organization that has traditionally opposed nuclear weapons, started getting calls from reporters, asking about isomer bombs. Many of the experts at the union had never heard of isomers, let alone an isomer bomb. Kurt Gottfried, a Cornell physics professor and cofounder of the union, began to investigate.

"Anyone with one course in physics can take hafnium-178, and on the back of an envelope, show that one ounce will heat a hundred tons of water from room temperature to boiling," he told me. "That sounds impressive, but that's a long way from doing it."

Gottfried admitted that he was not an expert on nuclear isomers, but said that after he got those first calls, he had looked into the matter, and it was clear there was a lot of hype. Along with the difficulty of storing a nuclear isomer weapon, there were some basic problems of quantum mechanics. For starters, he said, Collins's results presumed that the dental X-rays could excite the hafnium isomer, kicking it up to a higher energy level, which was followed by a prompt decay, sort of like pushing something up to the top rung of a ladder so that it falls over the edge. But according to Gottfried, physical laws demonstrated that this metaphorical "top rung on the ladder" didn't exist, or at least not where Collins claimed it could be.

"I think this is very farfetched," Gottfried said.

He had put together a brief information paper, explaining the reasons why the gamma-ray bomb wasn't a likely threat. That hadn't made much of an impression on some journalists, who preferred instead to trumpet the isomer arms race.

There were essentially two ways to write the story. An article could discuss the scientific controversy behind the bomb, which was an interesting but hardly sensational story. Or, option two was to write a breathless article about a futuristic nuclear weapon that would spark a new arms race. *The San Francisco Chronicle* took the former approach, and in late September 2003, ran a front-page article recounting William Herrmannsfeldt's battle against the isomer bomb. Less than two weeks later, *New Scientist*, a British general science publication known for its exposés of secret weapons, took a different tack.

"Gamma Ray Weapons Could Trigger Next Arms Race"

read the magazine's breathless August 2003 headline. *New Scientist* even put out a prepublication press release to promote their big scoop. Within days of its release, the *New Scientist* article spawned hundreds of Internet citations and was posted on a wide array of chat sites. The article was short on evidence, though, unsurprisingly, it quoted Carl Collins at length. The news also sparked a minor outcry in Europe. Articles in the U.K.'s *Guardian*, and in Germany's leading weekly magazine, *Der Spiegel*, and a daily newspaper, *Frankfurter Rundschau*, quickly followed.

The independent authority quoted in many of those articles was not one of the many scientists who had actually worked on the isomer research. Instead, the resident expert on the isomer bomb was a physicist named Andre Gsponer, who was frequently quoted warning that nations would be helpless at the hands of a country that possessed the fearsome isomer bomb. Maybe for some journalists, one nuclear weapons expert was as good as the next, because Gsponer proved to be a popular source for quotes. Yet Gsponer was neither a scientist associated with any noted university, nor a research physicist, a quick Google search revealed.

Who was Andre Gsponer?

None of the other scientists I had spoken with had ever heard of Gsponer's Swiss-based "Independent Scientific Research Institute," although Gsponer claimed it had been around for about twenty years.

"I will say little over the phone, we are low profile," Gsponer told me conspiratorially when I reached him at his office in Switzerland. "We are doing what you would call 'consultation work' on advanced technology—we work for private institutions and industry."

He didn't name any clients, however.

Gsponer's self-published papers littered the Internet, including such weighty pieces as "Antimatter Underestimated," which warned of an impending arms race in antimatter weapons. His investigations typically covered allegedly secret efforts to build new nuclear weapons, and of course, isomer weapons. All this proved perfect credentials, it turned out, to comment on the hafnium bomb. Gsponer was very excited to talk about isomers, which he was convinced were part of a "black," or secret program to develop a new class of weapons. Gsponer followed all these secret weapons, he told me, because his main obsession over the past twenty years had been fourth-generation nuclear weapons (the first being fission, the second fusion, and the third, the nuclear-pumped X-ray laser). Everywhere Gsponer looked, he saw secret fourth-generation nuclear weapons. The National Ignition Facility, a massive science project housed at the Livermore lab in California, was a secret fourth-generation nuclear weapons project. So, too, was ITER, the ambitious international fusion project ostensibly designed to produce energy. That's where isomers came in, he said.

Although Gsponer's knowledge of isomer weapons didn't seem to extend much beyond press accounts, he insisted that the idea of an "isomer bomb"—based on pure isomer—was not the real goal at all. Rather, the U.S. government wanted to use a small amount of isomer to trigger a fusion bomb, effectively replacing the fission primary currently used on such weapons. Then you could have mini-fusion bombs, long the stuff of science fiction.

Such a concept—of using isomers to replace the fission trigger used in thermonuclear weapons—had once been considered, some nuclear weapons scientists later told me, but never as more than a passing suggestion. But Gsponer was convinced that was the *only* purpose of isomer research.

"With isomers now, we are possibly in the phase of building a dedicated facility in the U.S. that is larger than those that exist already," he told me. "You have people like Collins and others who hope to be the main participants in that, so they are doing their lobbying."

Gsponer offered no proof of this secret isomer facility. How could you prove it? After all, it was secret. He did acknowledge that as a physicist, even he found Collins's results with hafnium quite unlikely. After all, it violated basic physics. But, he suggested, the U.S. government would eventually find another isomer, besides hafnium, that would work. That, presumably, was also based on his intuition, because he likewise offered no proof.

But for all the fears of Russian isomer weapons, no one seemed to be checking what the Russians were actually saying about hafnium. If they had, they might have found an article called, "New Generation of Nuclear Weapons and their Radiation Health Aspects," published in the January 2004 edition of the Russian-language *Moscow Bulletin on Atomic Energy,* and translated by the CIA-funded Foreign Broadcast Information Service.

The Russian scientists wrote of the great attention being paid to the hafnium controversy, particularly because of its possible application in the United States' war on terrorism.

"This would be a weapon comparable in effectiveness to low-yield nuclear warheads that would not cause radioactive contamination, would not result in wide-scale physical destruction, and most importantly would not fall under any international restrictions," the paper said.

According to some information, the Russian paper added, "the administration has already allocated two billion dollars for research in this field."

Since no one had at that point bothered to look at the health effects of such a bomb, the Russian scientists decided it was time to point out a few common sense things that the Pentagon hadn't considered. "Upon explosion of the warhead, hafnium would be scattered, contaminating the environment," the scientists noted.

It turns out, the toxic effects of hafnium had been studied on animals. In high enough doses, hafnium could cause a host of pulmonary problems, even leading to death. Of course, that didn't mean the Russians were rushing out to build a hafnium bomb. In fact, they were downright dubious about the whole venture.

"The situation as a whole is analogous to the story of cold fusion that came to nothing. And although situations like this are not rare in science, only a few of them have attracted widespread attention in virtue of the great consequences of the presumed successful scientific realization of the proposed idea."

Hafnium, it seemed, wasn't fooling the Russians.

12

FRINGE SCIENCE TAKES FLIGHT

THERE IS NO red flag, no marking point, nor clear sign that will say when a particular scientific endeavor crosses that line from dubious to silly. At one point do government officials cross over that invisible threshold and find themselves deep in the scientific underworld? Maybe there's no way to know, but hafnium supporters were coming up with ever more elaborate reasons why other experiments couldn't find triggering—or in some cases, why even Carl Collins's experiments weren't really reproducing. One scientist dubbed this phenomenon the "bad photons day" excuse, while another compared it to the alchemists' claim that the "the rooster whose blood was mixed in was not a pure black one." Whatever the case, hafnium was moving dangerously close to the gateway of the scientific underworld and all the warning signs were there.

Perhaps one sign that DARPA might have picked up on would have been Collins's increasing claims of a vast conspiracy of the scientific elite working to suppress his groundbreaking work. As he wrote in what became one in a long string of e-mail tirades he would send me, railing against the scientific establishment:

> In the U.S. we have largely a zero-base research budget. All support must come from a few sources that resort to external review of the proposals. It sounds good on paper, but what happens is that a

few people with very little time and many heavy responsibilities get a lot of proposals to score for support and the human condition insures [sic] that "Name" scientists and "Name" institutions get favored. Then the system tends to get defensive projecting the image of infallibility.

Who was driving this scientific elitism in Collins's mind? "The very hearts of the Bureaus of Innovation are the DOE labs at Livermore, Los Alamos, and Argonne," he wrote.

Of course, if Collins's belief in the ubiquitous control of the "Bureaus of Innovations" were true, it didn't explain why DARPA was dismissing the advice from those very laboratories that allegedly controlled scientific research. Rather, DARPA seemed quite happy to take the word of a Texas scientist, whose work was rejected by peer review, and whose rather generous funding came straight from the Pentagon.

How close did DARPA think it actually was to a weapon? After all, even Paul Robinson, the director of the Sandia lab who also supported the hafnium program, had told me that applications were likely to be decades away. Since DARPA wasn't answering any of my questions about the hafnium bomb, I decided to track down Tony Tether, the agency director, as he was leaving a military conference in Washington.

"Excuse me, Dr. Tether," I began, firmly placing myself between the DARPA director and his assistant.

Tether smiled.

"I wanted to ask you about hafnium," I continued. "How long would it be before hafnium would be a weapon? Would it be decades?"

"Decades?" Tether said, with a note of genuine surprise. "No, no. If the whole thing works, it's probably about ten years away," he said.

Tether's assistant stepped in. "We've got to go," he said.

Tether smiled and waved goodbye.

A month later, Tether followed up with a brief two-page statement sent to me through his spokeswoman. In a letter that ranged from the hyperbolic to the ridiculous, Tether warned of suicide bombers and Russian isomer weapons.

"What technologies could allow an adversary to unleash a technological surprise on us?" Tether asked, and "what technologies are needed to ensure our forces will always be able to unleash a decisive and deadly technological surprise on adversaries?"

The answer to both could be the hafnium bomb, Tether said.

> An enemy with this capability could create havoc on a scale that has never been seen before. For example, a car loaded with 100 lbs of Isomer could have the explosive power of 25 to 50 tons of TNT, or alternatively, a suicide bomber with a few pounds of Isomer could have the explosive power of a ton of TNT.
>
> We know that there are other countries, such as the Former Soviet Union, interested in Isomer weapons. Our research effort will help us answer the question of whether this is a threat that we need to worry about.
>
> On the other hand, having the capability ourselves would give the U.S. a capability that would truly be revolutionary given our ability to deliver small munitions with incredible precision.

Tether painted a fearsome portrait of the dangers posed by a hafnium bomb. American cities would live in terror of hafnium suicide bombers, the military would fear rogue states possessing hafnium bombs, and a resurgent Russia could threaten the United States with a new generation of nuclear arms. Of course, there was one other reason for the United States to spend money on hafnium, Tether pointed out.

"The U.S. could use this capability as a deterrent."

IT WAS A rainy morning in late February 2004, and William Herrmannsfeldt was still asleep at his home in Los Altos, California when Tom Ward called him about DARPA's isomer program. Ward, who used to work in the National Nuclear Security Administration, was now a consultant for a private company called TechSource. Like many of the odd business connections in the isomer program, no one knew for sure if Ward still held a government position, but TechSource was definitely on the DARPA payroll and Ward, who had made the mistake of bringing Herrmannsfeldt onto the production panel, had a financial stake in the program. Ward was upset, and he had big news about hafnium.

According to Ward, Johnny Foster and John Nuckles, two former directors of the Livermore lab, had picked up the phone to make a joint call to DARPA's director, Tony Tether, to complain about the isomer bomb. Foster and Nuckles were heavyweights in the world of government science, and it was a bit unusual, if not precedent-setting, for two of the country's top weapons scientists to call a Pentagon official to complain about a science program. But isomers had hit a raw nerve, and worse, Foster and Nuckles felt that DARPA was making a fool of itself at the expense of the Livermore lab. DARPA, after all, was investing in a weapon based on a dental X-ray experiment, and the agency was treating that work as more important than the best experimental scientists at the nuclear laboratories, who were calling the program a sham. It was too much for the former lab directors. DARPA could be wild and crazy, but not if it meant making everyone else look crazy in the process.

It was also a turf issue. The National Nuclear Security Administration, which operates the nuclear labs, was supposed to have purview over any nuclear energy programs, including weapons. So not only was DARPA supporting an idea gone wild, but by doing so, it was also helping the Pentagon encroach on what would normally be the territory of the national laboratories.

"Anti-lobbies count when it's from people like Johnny Foster who can easily destroy your career if you are where Tether sits," Herrmannsfeldt pointed out in an e-mail to me recalling news of Tether's encounter with the two former lab directors.

It had been almost two years since the isomer project had started, DARPA was supposed to begin producing hafnium soon for a bomb experiment, and the nation's top nuclear weapon scientists were kicking up a fuss.

Tether relented, a bit. He wouldn't cancel the program, he told them, but he would limit the funding to isomer triggering experiments. Tether decided DARPA would hold off on production plans until everyone was convinced hafnium triggered and he slashed the two-year budget from $30 million to just $7 million. It was an easy decision in a year when DARPA's overall budget was under severe pressure, but it was still a lot of money.

In the budget submission to Congress, DARPA also scrubbed the language describing the hafnium bomb. Gone was talk of an isomer weapon, leaving just a dull description of the enormous energy potential of hafnium.

"The goal is to scientifically prove, to the satisfaction of the majority of the nuclear physics community, that the stimulated isomer energy-release phenomenon does exist, that it can be reliably triggered by X-rays, and that the energy released justifies continuation of the program," Jan Walker, Tether's spokeswoman, said of the funding cut. She didn't mention Nuckles and Foster's phone call to Tether.

Considering that no one in the nuclear physics community—outside of Collins's group—believed that hafnium could be triggered with X-rays, it was curious what sort of proof DARPA planned on coming up with that would

convince a "majority" of nuclear physicists that the results
were legitimate. Was Tether planning to line them up to
observe a hafnium triggering experiment?

There was also a problem. Martin Stickley, the DARPA
program manager, had final say over how to divvy up the
rest of the funds. The choice was easy. Stickley immediately
cut off support for James Carroll, whose experiments were
finding only null results. When you only have $7 million
left in funding, there's no room to fund the doubters. The
remaining money was to be divided up between Carl Collins and Patrick McDaniel, the Sandia researcher who was
Collins's closest ally.

Apparently, Ward, the consultant from TechSource,
was upset by the funding cut, but he was also worried that
Stickley had essentially eliminated any scientific rigor. Ward
supported the program and had his own theory about how
triggering would work, but he recognized Collins as a threat
to any legitimate scientific work. He had argued with Ehsan
Khan to at least keep Carroll involved in hafnium, but Khan
was a Department of Energy official and just an advisor to
the program. Stickley was the manager and he wanted to
stick with McDaniel and Collins.

Ward didn't have much time to talk with Herrmannsfeldt. He and Khan were trying to save the hafnium bomb,
and to do that, they needed to assuage the critics. They were
scheduled to meet with Nuckles at Livermore that day. They
also planned to meet with Lowell Wood, another Livermore
scientist, who had close connections to the laboratory's leadership. Wood himself has a flair for some outlandish science;
his reputation came from promoting the Star Wars–era X-ray
laser, and also for conceiving the follow-on, Brilliant Pebbles,
space-based missile interceptors that never got much beyond
briefing slides. In later years, he had made his way into the
world of cold fusion. Perhaps for that reason, he seemed a
likely candidate to support the hafnium bomb. If Khan and

Ward could convince laboratory denizens like Wood and Nuckles that isomer triggering had some legitimate science, there was a chance to salvage things.

Khan and Ward also spent over two hours at Livermore speaking with an extremely skeptical Mort Weiss. Weiss, who had argued against the program from the start, had a rather perplexed account of the meeting.

"I will say that a number of serious questions were asked and those who asked them were not told to get lost," Weiss reported dryly.

But what seemed surprising, according to Weiss, was that a lot of money had been spent without getting any clear answers on basic issues of science.

Khan and Ward didn't make much headway with the Livermore crowd. In fact, they would concede that the experimental results over the past year had failed to provide clear evidence of triggering. Collins was proving uncooperative and was unwilling to repeat his original 1998 dental X-ray experiment. Worse, the experimental results over the past year had too much background noise, a high margin of error, and other problems. The past year's experiments, Stickley admitted in response to a Congressional inquiry, had not been as successful as he had hoped.

But if physicists were complaining that DARPA's hafnium research was embarrassing the government, they didn't know the half of it. The same week that Nuckles and Foster called Tether, Llew Smith, a Welsh Labor member in the British Parliament and gadfly on nuclear issues, submitted a written question to Geoffrey Hoon, the U.K. minister of defense. Smith wanted to know if the United Kingdom was cooperating with "Hafnium-178 research at the U.S. Air Force Research Laboratory at Kirtland in New Mexico" and demanded to know what research the United Kingdom's own nuclear weapons establishment was doing.

In a written reply, Hoon assured Smith that the

United Kingdom was not cooperating on any research into hafnium-178.

Hafnium hysteria had reached new heights. From Romania to Texas to the Pentagon and now over to the United Kingdom, the isomer bomb had liberal politicians in Europe worried about a new arms race.

IT WAS A month after the Nuckles and Foster call, and Tether was having a rough day when I ran into him at a congressional hearing in March 2004. Tether looked nervous—not an unusual reaction for Pentagon officials about to testify in front of the House Armed Services Committee. It was budget time on Capitol Hill and Tether had more than just the usual grilling ahead of him. The week before the hearing, DARPA's latest publicity stunt, the DARPA Grand Challenge, had not gone well. Or more accurately, it hadn't gone very far.

The ambitious event was supposed to be the hallmark of the agency's innovative methods. DARPA offered a $1 million cash jackpot to the winner of a robot race that would weave through the Mojave Dessert. The goal was to race the unmanned vehicles—a collection of converted trucks, jury-rigged cars, and other ungainly vehicles, to the finish line without any human intervention. The race was even more unusual because unlike a typical military project, the competition was open to anyone and everyone, from the itinerant garage inventor to high-school students and experienced engineers.

After over a year of good publicity lauding DARPA's innovative effort to push the envelope on technology, the much-vaunted race ended after just seven miles, when the last of the unmanned trucks and vehicles crashed and burned.

"Robot Race Suffers Quick, Ignoble End," jeered one newspaper headline. "Pentagon's Robot Race Stalled at Gate," chimed another.

Tether also knew that the *Washington Post Magazine* was

about to run a long cover story I had written on the hafnium bomb—the timing couldn't be worse. DARPA was trying to shake the image of being out of control and yet its projects were being openly mocked.

Sitting in the hearing room that day, I also saw Ronald Sega, the Pentagon's chief technology official. He waved at me and smiled. The phrasealator was nowhere in sight.

The lawmakers were running late and Tether stopped on the way to the witness table to shake my hand and chat for a minute. In his usual friendly manner, he asked whether I had gotten "all my hafnium questions answered."

I pointed out to him that DARPA, as he was aware, had refused to answer almost all of my questions in the five months I had attempted to get information about the program. In fact, with the exception of the brief essay that Tether had penned to me about "isomer suicide bombers," DARPA had refused to communicate with me in any way. Tether shrugged his shoulders sheepishly, but didn't offer any explanation.

"The article's fair," I offered.

"Well, if it's not, we know where you live," Tether joked with a nervous smile, before ambling over to his seat.

It suddenly occurred to me that it wasn't such a funny joke for an agency that runs projects with Orwellian names like "Cities that See."

Over the course of the next two months, hafnium was about to get more publicity than Tether ever expected a basic science program might attract.

Engineers have a difficult time understanding the bitter disputes of physicists. Helicopters fly or they don't, and hafnium triggers or it doesn't. Or at least this seemed to be Tether's understanding of the situation.

"There is too great a fuss over what seems to be a straightforward thing to do and find out the answer to, one way or another," Tether had vented in frustration at one point.

If triggering hafnium was as easy as shining a fifteen-hundred-dollar dental X-ray machine on a target, then how hard could it be to prove that to a bunch of skeptical physicists?

Between William Herrmannsfeldt's battle to kill the hafnium project and a growing wave of negative publicity, Tether decided that there had to be one simple way to solve the issue: conduct a definitive experiment that would determine, once and for all, if hafnium triggered. More important, it would be an experiment public enough to silence the critics. The engineer's world works through demonstration and Tether thought that if Stickley could demonstrate hafnium triggering in a semi-public forum, everyone would have to believe it. In what was a highly unusual move, Tether ordered Stickley to send a letter out to all of the scientists involved in the isomer debate, particularly the critics. The letter explained that DARPA was going to design a definitive experiment. They had even come up with a special acronym. The experiment was dubbed the TRiggered Isomer Proof of Principle, or TRIP, and the agency wanted input from everyone, even the critics, on how to design the TRIP experiment.

There was a catch. DARPA expected the critics to agree to the final results.

"If you agree to the plan and triggering does not occur, then we will conclude that the hafnium isomer does not trigger," Stickley wrote in a letter, which Tether himself edited. "On the other hand, if you agree to the plan and it does trigger, then, conversely, we will conclude that the hafnium isomer can be triggered with X-rays."

Then there was another issue, although not one that Stickley or Tether had put in the letter. Patrick McDaniel, Collins's closest ally, would conduct the definitive experiment.

But even under these favorable conditions, Collins was furious. And in his usual response to perceived attacks, he sat down to write a slew of e-mails, sending copies to his supporters, such as Nancy Ries and McDaniel, but also to

Paul Robinson, the head of Sandia. There is no such thing as a test that can disprove isomer triggering, he insisted. "The ONLY DEFINITIVE TEST is whether others can reproduce that with positive results," Collins wrote. "Just do the same things."

Of course "the same things" were the problem. Every time someone repeated the triggering experiment and came up with a null result, Collins would claim they had gotten some essential detail of the experimental setup wrong. And each "detail" would happen to be something not included in his original paper.

First it was the wrong triggering level, then it was the wrong type of X-rays, and then the measurements were being done at the wrong time. And, of course, Collins had never revealed to anyone the exact composition of his precious hafnium target, although he accused others of not preparing their hafnium correctly. Collins's articles would typically say only how many hafnium nuclei were in the target, and that it was encased in plastic.

In the case of Argonne, the scientists prepared the hafnium in oxidized, or powdered, form, and encased it between two very thin layers of beryllium metal foil. Plastic, they reasoned, might swell and crack under X-rays. "Since the powder is radioactive, you have to maintain the integrity of the casing or you can find yourself cleaning up radioactive powder that spreads rapidly around, given a chance," said Donald Gemmell, one of the Argonne physicists. "That would be expensive and would sharply diminish your popularity amongst your colleagues at the synchrotron."

Collins's response: they constructed their target wrong. And so it went.

Worse, Collins's writings were becoming increasingly irate, turning into a garbled mix of odd syntax and idiosyncratic capitalization. In one e-mail blasting DARPA for its proposed counter-experiment, Collins wrote:

The course proposed is egregiously flawed, logically and legally. It is the intellectual equivalent of a lynching. Being proposed is a sort of political palliative aimed to concoct an "equivalent" experiment, purported to be so wisely configured that its success may (or may not) still the passions of the critics, but its failure FOR ANY REASON including non equivalence will surely destroy the careers, futures, financial support, and equities belonging to a wide cohort of Students, Faculty, an Institutions all of whom had the courage, vision, and effort to give 100% during the last 15 years of success. You cannot conclude from a failed equivalence that "isomer triggering does not occur," because it does. What could be concluded from a failed equivalency is that the courage of the Sponsors is insufficient to justify continuing support within the political context created by the opponents.

Collins also typed out another long e-mail to Stickley, warning again that the "High Priests of Innovation" were out to destroy his work. After denying, time and time again, that he had given any thought to the idea of weapons research, Collins suddenly grasped on to the threat of rogue isomer bombs as the *raison d'etre* for the program. The Russian lab in Dubna, which had produced the world's purest isomer sample, was also a training facility for North Korean weapons scientists, Collins warned.

A hafnium bomb was close at hand, reported Collins. By his calculations, a seventy-five-milligram sample of hafnium, or "one-fifth of an aspirin tablet" would be all the hafnium needed to get a "critical mass" for a bomb. "What it shows is that a nuclear 9/11 is perfectly possible if the material can be made; patiently in small amounts," Collins wrote.

"God forbid, if there is a nuclear 9/11, my story is clear," Collins wrote to Stickley with a touch of hysteria. "I did my best against terrific odds to sound the warning of technical breakout. So, what are you going to do?"

The risk of a Russian or North Korean isomer bomb seemed a bit unlikely, however, if hafnium triggering didn't work. When I asked Philip Coyle, the former associate director of the Livermore lab, who also once served as the Pentagon's top technology tester, about North Korea's isomer program, he started laughing. "If you're worried about proliferation, you probably want to encourage a country like North Korea to work on isomer bombs."

It would be wonderful if North Korea started investing in isomer weapons, Coyle continued. "I can't think of anything better than for them to spend all of their money on the isomer bomb. They probably won't, but if they did, it would keep them from working on nuclear weapons that would actually work."

Collins, in the meantime, continued his tirade. As an alternative to the "draconian" test proposed in the DARPA letter, Collins submitted to Stickley an alterative. Turning his sights on his former student, Collins argued that if James Carroll would accept a definitive "triggering line" that could be reproduced similar to the original dental X-ray experiment, then that should be the accepted proof. Why Collins was so focused on Carroll, when it was the nuclear physics community opposing his results, was unclear. But Collins's counterproposal was barely coherent, sounding at times dangerously close to the ramblings of the autodidacts who write from prison cells with grand unification theories of the universe.

"Current concerns over the challenges to this established result seem to arise from political agitation over the possible consequences of this discovery to Society," Collins wrote in his alternative proposal.

In a more unusual e-mail between a government official and a sponsored scientist, Stickley replied to Collins with assurances that plans for a definitive experiment—something Stickley himself had opposed—would quickly prove the Texas scientist was right. Writing a hurried e-mail from his hand-held Blackberry, the chief patron of isomer triggering wrote an almost touching note to his primary beneficiary.

"Carl—it is because of the quality of your work and that of your colleagues and students that this will be successful," Stickley assured Collins. "I am convinced of that."

Stickley went on to assure Collins that McDaniel, with Collins's help, would have no problem duplicating the dental X-ray results. In an even more surprising statement Stickley went on to say: "It is clear that the latest data you have taken that trigering [sic] is real." A few days later, Nancy Ries, the Sandia official in charge of the isomer program and longtime friend of McDaniel, wrote me an e-mail out of the blue. She had never answered my initial request to speak to her about isomers, so I was surprised months later when her name appeared in my inbox. Even more surprising was that, like Stickley's note to Collins, Ries's e-mail had an almost inno-cent, childlike quality. She wrote about how hard it was to move the isomer program forward with "adversaries" work-ing against it.

"Carl is a world class scientist and human being as well," she wrote, adding one more unusual sentence for a government official involved in the nuclear weapons busi-ness: "He has such a good heart and I hope will be my friend forever."

The "hafnium bomb" was showing up on Web sites like www.patheticearthlings.com and the hafnium gap had become a one-line joke in the White House's Office of Sci-ence and Technology Policy. But the believers still believed hafnium would trigger and nothing, it seemed, was going to shake them from that conviction.

The refusal to listen to the near-deafening roar of independent scientific advice was almost incomprehensible. After six months, I had wrestled with every theory imaginable that could explain why the hafnium believers refused to give up their increasingly silly belief in triggering.

Writing from his home near Livermore, California, Weiss had a rather cynical explanation.

"Scientists can be the servants of politicians but not science and somehow that appears not to be understood within the Beltway. Helicopters crash because of turbulence and Hafnium doesn't trigger or chain and can't be made [to] because nuclear physics is not subject to political whimsy."

In Weiss's view, DARPA should have responded to the original JASON study by asking for an independent review, similar to what was done for cold fusion. Instead, he noted, the agency chose simply to ignore the JASONs, Herrmannsfeldt's request for a review, and "theoretical arguments that would have made an ox skeptical." This is a trend characteristic of more than DARPA, Weiss lamented, and reflects a larger problem in government and science.

"Choosing one's advice from sycophants is an ancient road to disaster," he concluded.

IN EARLY APRIL 2004, Esen Alp, a senior physicist at the Argonne lab, appeared in the office of the director of the SPring-8 synchrotron in Japan, where Collins was conducting his post–dental X-ray experiments. Alp was another in a long line of Collins critics who had bristled when the obscure Texas scientist carried on about the "failures" at the Argonne lab. Alp was at the Japanese facility for routine work, but he brought with him one small present for the director at SPring-8, a copy of my *Washington Post Magazine* article featuring the cover story on Carl Collins and DARPA's nuclear hand grenade. He wanted to be sure the director got a copy of the article. Pictures and all.

The next day, Collins received an e-mail from Yoshitake Yoda, his main Japanese collaborator at SPring-8. Subject line: "I am sorry the crisis in collaboration."

In polite if somewhat grammatically challenged English, the Japanese scientist explained to Collins the preceding week's events, which had involved a rather unpleasant conversation between him and the director.

"I do not care the contraposition to other groups," Yoda told Collins in the sorrowful e-mail recounting the exchange. "It is really a scientific issue although my boss's feeling is a little bit different."

The problem, Yoda continued, was not the science but the nuclear hand grenade. SPring-8 was a government facility and operated under a strict mandate to promote peaceful scientific research and not weapons. Because of their use against Japanese cities in World War II, nuclear weapons are still regarded as anathema in Japan. The staff of SPring-8 had thought that Collins was only doing basic physics research, Yoda said, and the director was upset that the work was in fact aimed at developing a new type of nuclear weapon. Yoda wasn't too happy, either.

"I believe almost all the nation do not want to support the development of the weapons of USA. Especially we hate atomic bomb," Yoda wrote to Collins. "So wherever essence is, there is no chance as long as your research is supported for the weapons development."

The e-mail concluded with an explanation that the Japanese scientist would not be willing to complete work on a new shield he had promised for Collins's upcoming experiment. And Collins's experiment scheduled at the facility for that summer would most likely be cancelled.

Collins was enraged. In an e-mail to Stickley, he called Alp, the Argonne scientist, "a senior manipulator," who was only "claiming to be a scientist."

Apparently, Collins also tried to brush up on Defense

Department contracting regulations. He came to the unlikely conclusion that, by endangering his work at SPring-8, the Argonne scientist's actions must be a crime under federal procurement law.

"To whom do we report that; or am I incorrect in understanding it to be a crime?" Collins asked Stickley.

JAMES CARROLL LOOKED as much out of place at the Ritz Carlton hotel in McLean, Virginia, as any Midwestern professor of physics could look. The wood-paneled lobby and darkened martini bar was filled with prosperous-looking businessmen whose work placed them in the higher income brackets of Washington. Carroll, dressed in a sweater and a pair of jeans, was in northern Virginia for not altogether clear purposes that involved his work with isomers; he had agreed to meet for a casual dinner. Looking a bit embarrassed, he said that the restaurants at the Ritz were perhaps a bit too stuffy for his tastes and we headed out of the hotel to the mall connected through the back of the lobby.

In the same way that Washington's midlevel government buildings mesh comfortably with swank office space, the galleria that connects to the Ritz makes an almost shocking transition from high-end luxury to middlebrow suburbia. We ended up at a chain restaurant and ordered some beers.

Friendly and soft-spoken, it was hard to imagine Carroll engaging in the sort of histrionics that Collins, his former advisor, preferred. But it was also easy to see how Collins, who stood over six feet tall and was a dynamic speaker, must have once towered over his young graduate student. Now in his forties, Carroll looked much like the slightly built student from the fading picture in the Texas lab.

Carroll was nonconfrontational and had an effortless way of steering conversations away from subjects he didn't want to discuss—topics like the CIA's support for his work. But if Collins feigned ignorance of issues he wanted to avoid (weapons, for instance), Carroll was honest and merely adept at politely changing the subject. Working with Collins for eight years down in Texas was one of those subjects that he easily skirted until finally, in a self-deprecating manner, he acknowledged he was avoiding the subject. He didn't like to talk about Texas because he preferred speaking about the science behind his work and not the personalities.

The brouhaha over isomer weapons had tired him, however, and he did admit that at one point, when he and Collins were presenting at many of the same Pentagon meetings, he hatched his own form of revenge. When Collins would show his briefing slide of the golf ball filled with isomer explosive, Carroll would show his own slide of a gag golf ball blowing up in someone's face. The stunt didn't do much to improve relations between the two men, but it gave Carroll at least some sense of satisfaction.

He did concede that, at one point, back when he was Collins's favorite graduate student, he had looked up to his advisor as almost a father figure. Collins, Carroll finally said, could be a very likeable and charismatic person.

The description was not unlike the multitude of ways I had heard Collins described. "Able to sell ice cubes to an Eskimo," one Pentagon official said of the charismatic Texan.

When I asked Carroll what it was that made Stickley, Ries, and McDaniel so sure of Collins's results in the face of so much criticism and scientific dispute, Carroll just shook his head. He began to talk about the Jonestown massacre, the mass suicide of over nine hundred followers of the Reverend Jim Jones. At some point, the isomer believers had simply put their faith in Collins and any attack on Collins, or on his work, was an attack on that basic faith. They were all fellow travelers too far down the same road to consider looking back.

Carroll's assessment may have sounded dramatic, except

Collins really was writing hysterical letters about a "nuclear 9/11" involving hafnium and his supporters were writing back with fawning declarations of undying friendship and belief in his work. Ries had written about how Carl was "responsible for literally saving the lives and futures of many young Romanians." Stickley, ignoring the reports of his own advisors, was "convinced" the definitive experiment would work.

Carroll, for his part, said he wasn't all that disappointed about being cut from the DARPA project. The whole personality dynamics had left him uneasy, and he was able to make up most of the funding from the Air Force. The publicity around the DARPA program worried Carroll, however, particularly if it ended up tarring all work on isomers. Basic physics research on isomers, he insisted, may not be exciting, but it was fundamentally good science. It was also his life's work.

Of his former mentor, Carroll just shook his head. Collins had long had delusions of grandeur, claiming that his work was worthy of the Nobel Prize. It didn't even matter anymore to Collins that he couldn't get published in mainstream scientific journals, Carroll suggested; Collins wasn't interested in engaging in scientific dialogue anymore. For years, Collins had become more and more confrontational, less willing to accept criticism, and more prone to extreme positions.

"I think he's finally lost all restraint," Carroll said.

IN THE MIDDLE of DARPA's planning for the final hafnium experiment, the May 2004 issue of *Popular Mechanics*, the glossy weekly for hobbyists and car enthusiasts, showed up on newsstands. Its science editor, Jim Wilson, had always been known to have a penchant for science fiction, and his writing often ranged from space aliens and alleged military cover-ups to top-secret aircraft. One of his better-known articles was an "exposé" of Area 51, the secret Air Force testing range that has been rumored over the years to house everything from stealth aircraft to alien technology.

Perhaps that helped explain his feature story, called "Atomic Planes." The cover depicted a small passenger plane with what looked like a middle-American family on board. The teenage son was at the helm.

It was an artistic rendition of the hafnium-fueled aircraft. In another picture inside the magazine, the X-ray machine was visible in a cutaway drawing, with the telltale yellow radiation sign on the side of the aircraft.

Set between articles on high-definition car radios and the latest in power tools, the *Popular Mechanics* article waxed enthusiastic about how the magazine had discovered secretive plans for a new hafnium-fueled military aircraft that would spur the next revolution in flight. "The new hafnium-fueled reactor emits so little radiation it could be easily integrated into civil airport operations," the article proclaimed of the isomer, whose thirty-one-year half-life makes it, in any quantities above the microscopic level, one of the world's most radioactive materials, and thousands of times more radioactive than the materials packed into nuclear weapons.

In a unique mix of science and science fiction, the article reported that an Air Force captain named Chris Hamilton had been working on a hafnium reactor at the Air Force Research Laboratory and that plans were underway to take a Global Hawk unmanned aerial vehicle, the military's most advanced spy drone, and install a hafnium reactor on it. No more fuel; the unmanned aircraft would fly on hafnium.

"Should an accident occur, there is less of an environmental concern than with fission," Wilson enthusiastically declared. "Hafnium-178 has a half-life of only 31 years compared to thousands of years for other reactor fuels."

Inverting the relationship between half-lives and

radioactivity (the shorter the half-life, the more radioactive the element), the author placidly asserted that an accident involving hafnium wouldn't be that dangerous.

Wilson also made casual mention of Carl Collins's "remarkable and unexpected" discovery that hafnium-178 could be triggered with a dental X-ray machine. In a conspiratorial tone, his article suggested that detractors at the Department of Energy were using its labs to quash talk of nuclear reactors.

It wasn't, in fact, the first time that Hamilton, the Air Force captain, had his work on hafnium featured in an article. *New Scientist,* a general science magazine, had written almost an identical article one year earlier, as had *Flight International,* a British aerospace publication.

Hamilton's work on hafnium was not part of any official project, but an outgrowth of his graduate work in aeronautics. He had written an entire master's thesis on hafnium-powered aircraft, and his mentor and sponsor for the work, as it turns out, was Patrick McDaniel. Admittedly ignorant of nuclear physics—and too busy to take the time to learn about the subject—Hamilton had been promoting hafnium-fueled aircraft at aeronautics conferences.

The *New Scientist* article, which repeated the bizarre assertion that hafnium was relatively safe to be around, quoted the Air Force captain at length. Hamilton had told people that the article was "misleading," although he declined to specify what exactly he disagreed with in the story. When I tracked Hamilton down at Edwards Air Force Base in California, he claimed that he had never spoken to the *Popular Mechanics* reporter, whose article also mysteriously quoted him at length. Hamilton surmised that the quotes were taken from the *New Scientist* article, or perhaps from a conference he presented at in Nevada. In either case, Hamilton hastily added he was in intense training and was too busy to speak with me by phone.

Writing by e-mail, Hamilton explained that he had never bothered to look at the physics. "My research did not prove or disprove the triggered isomer phenomenon but was a study of one application if it was possible to use hafnium as a power source," he wrote. He promised to write again and answer more questions, but after offering a few plaintive excuses involving sick relatives, that was the last I heard from him.

Hamilton, in fact, was one of two graduate students Patrick McDaniel had encouraged to pursue work on hafnium reactors. McDaniel, whose main preoccupation was with isomer weapons, had suddenly found a keen interest in encouraging young military students to study hafnium-powered aircraft and had arranged for the Air Force to provide some funding for their research. If funding gullible graduate students to dream about hafnium-fueled aircraft wasn't a strategic plan to present a friendly public face for the hafnium bomb, it certainly looked that way.

The hafnium aircraft article was enough to catch the attention of Robert Park, the head of public information for the American Physical Society. Park, who writes a weekly e-mail newsletter called, "What's New," had made a second career watching what he calls "voodoo science." He put the *Popular Mechanics* article at the top of his weekly column, with the heading: "Just when you thought life couldn't get any sillier."

Drawing attention to the article on hafnium-fueled reactors, Park poked fun at the concept by comparing it to the 1950s-era concept for a nuclear-powered aircraft, a sort of flying Three Mile Island. "The problem with fission reactors was that they required too much shielding," Park wrote. "The problem with the hafnium-178 reactor is that it doesn't exist."

With the *Popular Mechanics* article, hafnium, Park concluded, was officially ready to join cold fusion, zero-point energy, and all the rest of the voodoo science that littered the yellow-bricked road leading to the netherworld of unlimited energy.

William Herrmannsfeldt, in the meantime, was crushed. He had been working for nearly two years to kill the isomer program and now a popular magazine was hailing it as the next revolution in flight.

"Friends . . . this stuff is hard to kill, worse than dandelions in the yard," he wrote to the various members of the Argonne group, whose experiment had contradicted Collins's dental X-ray results.

But things were about to get worse for Collins. Bert Schwarzschild, a reporter for *Physics Today*, decided it was time for the stalwart publication of the physics community to take a crack at the hafnium debate. Schwarzschild noted the recent cut to the isomer budget, but quickly declared that still left the project with "too much money." He was also about to interview Carl Collins.

The interview did not go well. According to Schwarzschild's later account, things went downhill rapidly when he began to question Collins about some of the published results disputing isomer triggering. Collins inevitably launched into his customary diatribe on the Argonne "failures" and Schwarzschild interjected to announce he was taping the conversation.

Collins exploded. In what seemed like an odd move for a physicist, Collins wrote directly wrote to Martin Blume, head of the American Physical Society, which runs, among other things, the *Physical Review* journals. Collins complained that Schwarzschild had "baited" him into making negative comments about Argonne and then "tricked" him with the phone recording. While Collins's account of the events was rambling and included the usual odd capitalization, what he recounted was more or less true.

Except that Martin Blume had nothing to do with *Physics Today*, which is part of the American Institute for Physics.

The mistake seemed odd for Collins, a professor of physics. An extremely polite, if amused, Blume replied to Collins to correct him. Collins subsequently called the editors of *Physics Today*, but still seemed confused. Just two weeks later Collins again referred to *Physics Today* in one of his mass e-mailings as "the popular organ of the most important professional society for Physics in the USA, namely the American Physical Society."

Schwarzschild was stunned by Collins's erratic behavior.

"Is he mentally ill?" Schwarzschild asked me after his run-in with Collins.

I thought about it for a moment. "No," I replied. "I don't think that's quite the problem."

The article that came out in *Physics Today*, was not particularly critical of Collins's experiment, but it didn't need to be. By quoting the heavyweights of physics in the popular journal of the scientific community, the article further marginalized Collins's work. And worse, Collins's bizarre letters addressed to "Editor Blume" of the American Physical Society had certainly pegged him as an oddball—and not in the endearing, absentminded-professor sort of way.

Collins didn't realize it, but his work had gone from *Physical Review*—the leading journal of the scientific community—to science fiction chat rooms.

But he was still getting money from DARPA.

As bad as things looked for the DARPA program, I couldn't help but feel a little sympathy for Collins. Looking down at my desk, I noticed an old e-mail, sent to me from Collins's "number one enemy," Donald Gemmell. Collins, who had never met Gemmell, had often called the Argonne physicist "vile," accusing him of being a "politician, not a scientist."

But Gemmell seemed to have a touch of empathy for Collins.

"It fascinates me, as I said before, to contemplate what it is that makes Collins 'tick,'" Gemmell had written to me one evening, trying for a moment to imagine what would it be like to be in Collins's shoes. "Trying day after day to use statistically inadequate data to persuade a bunch of skeptical physicists that he has made a great breakthrough that will open the way into gamma-ray lasers, new weapons, controllable triggerable releases of enormous energy, etc. He must be persuasive and he must enjoy the limelight. But he must worry about the inevitable day of reckoning."

Collins's day of reckoning was perhaps finally on its way.

13

WELCOME TO THE FAR SIDE

THERE WAS NO better sign of the times in Washington than a sudden announcement in March 2004 that the Department of Energy, after fifteen years, was going to take another look at cold fusion. It would be the second time, in fact, that the federal government had assembled a panel of scientists to review the claims that a potentially cheap and unlimited source of energy was within the grasp of modern science.

In 1989, shortly after the initial cold fusion claims were made in Utah, the Department of Energy had assembled a panel of experts to review the experimental evidence. After visiting a variety of labs conducting cold fusion work, the panel publicly concluded there appeared to be no convincing evidence of room temperature nuclear reactions. Individually, the panel's members were also known for calling cold fusion's discoverers "incompetent boobs," "delusional," and similarly unflattering adjectives.

Why then, after so many years, was the Department of Energy again reviewing claims of cold fusion?

The ostensible reason for revisiting the topic sounded quite reasonable. After over a decade, several hundred researchers working on cold fusion were claiming better results. Peter Hagelstein, father of the never-built X-ray laser, had written a letter to the Department of Energy requesting the new cold fusion review. And James Decker, the senior Department of Energy official who had declined to pull out of the hafnium program, declared that it was his "personal judgment" that a new review of cold fusion was a reasonable request, made by scientists "from excellent scientific institutions" and possessing "excellent credentials."

Decker omitted mention of one other relevant fact. Even those in the cold fusion field acknowledged there was really only one reason why the government had agreed to a new review, and that reason was named Randall Hekman, a former judge turned Christian radio minister, turned cold-fusion enthusiast. Hekman had two things going for him, he told me when I contacted him: he knew the energy secretary, Spencer Abraham, and he knew cold fusion.

Hekman knew the energy secretary from his days in Michigan politics, and he knew cold fusion, because after Hekman retired as head of the Michigan Family Forum, a conservative Christian organization, he started working on cold fusion at home. Apparently, in Washington, those facts were enough to move scientific mountains, because soon enough, the Department of Energy agreed to a new review.

Cold fusion was not without its supporters in official Washington, particularly in the military. The Navy's research community was known for taking a sympathetic view of cold fusion, occasionally funding work in the area—or at least tolerating naval researchers who pursued cold fusion research. In 2004, when I asked the Office of Naval Research why they had continued to support cold fusion for so many years after the mainstream scientific community rejected it as pseudoscience, they provided me with an answer that wasn't quite at the gateway to the scientific underworld, but it was getting awfully close.

> The Navy and the defense of our country cannot afford to blindly accept the status quo. When science is delegated to a narrow conventional and "normal" view of what is possible, we severely limit our ability

to make large technological advances. Large advances
are made in two major ways, slow steady progress
of existing knowledge, and breakthrough concepts
that challenge the scientific status quo. Both are
valid. The United States enjoys leadership in the
scientific and technical community because of the
diversity of approaches we use to support research
ranging from the private sector through the varied
approaches of government funding agencies. We
should all fear the day when there is a "normal"
channel by which all new scientific ideas are evalu-
ated for support.

Does challenging the status quo mean challenging the
scientific method, or the notion of peer review? It was getting
increasingly hard to tell where that fuzzy line between high
risk, high payoff, and foolish risk, no payoff was located.

As if to test precisely the limits, the Air Force Research
Laboratory in August 2004 released a study it had funded
that at first glance, almost sounded scientific. But not quite.
The "Teleportation Physics Study," conducted by a company
called Warp Drive Metrics in Las Vegas, Nevada, proposed
to look at means of psychic conveyance. Or in other words,
teleportation, á la *Star Trek*.

"This study also consisted of a search for teleportation
phenomena occurring naturally or under laboratory conditions that can be assembled into a model describing the conditions required to accomplish the transfer of objects. This
included a review and documentation of quantum teleportation, its theoretical basis, technological development, and its
potential applications," the first page of the study read.

What exactly those "conditions" were—natural or unnat-
ural—were somewhat shrouded in mystery, however, because
reporters initially weren't able to contact the lead researcher,
identified as an Eric W. Davis of Las Vegas, Nevada.

Davis's treatise would have to speak for itself. In it, he sprinkled Einsteinian formulations and various physics equations with a mix of science fiction references and concepts ranging from telekinesis to parallel universes.

An Air Force spokesman explained the teleportation study to a reporter from *USA Today* by saying, "If we don't turn over stones, we don't know if we have missed something."

When Robert Park, the cantankerous spokesman for the American Physical Society, called the Air Force to ask how much the government had spent on the scientific equivalent of "beam me up, Scottie," he was told the military couldn't tell him how much the study cost (press reports put it at $25,000), because it was part of a much larger contract. Park did find out, however, that the Air Force Research Laboratory manager in charge of the study had a patent for "zero-point energy," the idea that unlimited energy could be tapped from the quantum vacuum.

I eventually found Davis through a circuitous route—Jack Sarfatti, inventor of the God phone—had included Davis on one of his massive e-mail chains. When I contacted Davis, he defended the Air Force's teleportation study, arguing that if the military didn't explore "off the wall" concepts, then other countries would eventually beat the United States to new discoveries.

"And then we will find ourselves in combat with our pants down when the enemy deploys a new weapon that renders our combat systems ineffective," he wrote.

But is every idea worthwhile to fund? Would spending money on psychics, teleportation, and perhaps isomer bombs, really yield a breakthrough? Davis argued that the military might fund a hundred offbeat concepts so that one would pay off. But what if none of them paid off? What if they were all bad ideas? After all, what ever did happen to Hal Puthoff, the man behind the CIA psychics program?

Davis himself answered that question.

"Anyway, since last fall I moved to Austin where I began anew working for my longtime friend and colleague, Dr. Hal Puthoff who directs the Inst. for Advanced Studies at Austin, whose research I cited many times in my Teleportation Physics Study."

The scientific underworld was thriving in Texas.

IT WAS THE spring of 2004, and if Tony Tether were having a hard day, you wouldn't know it. He was back at Disneyland at DARPA's annual convention, the hafnium project was in full swing, and speaking to a sold-out audience of technologists, the good-natured director was almost giddy with enthusiasm. Following a ten-minute video featuring DARPA program managers dressed in basketball uniforms, Tether was called on stage like the main act at a rock concert, replete with flashing lights and pulsing music.

"Welcome to the Far Side!" Tether boomed. "A world where science fiction morphs into reality."

What is the far side and who resides there? This was how Tony Tether, a man who controlled approximately $3 billion in annual funding, explained his mission:

> The people who inhabit the Far Side are anything but conventional. I know some of these people. So do you. Some of them are here in this audience. In fact, one of them may even be you. You know who you are. You are always coming up with ideas, many of which do not prove out but then again some prove to be truly revolutionary. You are the one who visualizes combining two existing systems to produce a capability that no one else had even thought possible. You are the one with the idea to manipulate atoms to atoms in a wonderful new material, a new material like, Un-a-tainium. No

doubt, you have been frustrated because no one on the Near-Side will fund your idea. But DARPA might.

DARPA, it seemed, would fund just about anything.

Back in Washington, however, DARPA was facing problems. In early May, Martin Stickley, via Ehsan Khan, sent out an odd message to the various members of HIPP, the hafnium isomer production panel.

"Ehsan and I are beginning to get inquiries about the HIPP activity and HIPP report," Stickley said. "My message to you is that you do not discuss this with anyone nor raise it as a topic to discuss with people who have no need to know."

The knowledge gained about hafnium production was "valuable to the country" and panel members must guard this knowledge from anyone outside the panel, Stickley warned. Khan also added in a personal message: "Martin expects every HIPP member to do his duty," he wrote, adding thanks for a "a job well done."

The inquires that were coming into DARPA, it turns out, were from the staff of congressional armed services committees, who were demanding that DARPA explain the isomer business. The publicity sparked by the article in the *Washington Post Magazine* had suddenly awakened lawmakers to the fact that the Pentagon was pursuing a new type of nuclear weapon outside the purview of Congress. They were also worried about the millions being spent on a program of dubious scientific value.

Stickley responded to the committees with an upbeat overview of the program, explaining that the panel had found that production costs would "fall in the range required to produce typical nuclear weapon materials, and that the isomer could probably be produced at a militarily useful rate."

He went on to concede, however, that the latest triggering experiments had been less than successful for a "variety of reasons." The answer to this dilemma, of course, was to continue funding. The final experiment, which Stickley thought would vindicate hafnium, was fast approaching.

In the meantime, the critics of the hafnium bomb, in response to Stickley's letter requesting feedback on the final experiment, were beginning to offer their thoughts.

"I would very much like to have a chance to comment on the design of the actual experiment and on the details of the data analysis," Peter Zimmerman wrote in a June 3, 2004 reply to Stickley.

Zimmerman offered an intriguing suggestion. Since the dispute over hafnium triggering centered on statistically marginal results, why not do something akin to a "blind experiment," using a computer program to shift between "beam on" and "beam off" modes? The experimenter wouldn't know, until after looking at and analyzing the data, which runs were which.

"Why?" Zimmerman explained. "So they don't diddle the equipment to produce signals when the isomer is in the beam and not to produce signals when it's gone."

Stickley's response was not enthusiastic. "In discussing the advantages of a computer-controlled stage with numerous physicists, the advantages did not seem as strong as you implied," Stickley wrote in a letter to Zimmerman, dated August 31.

In other words, Stickley didn't see the need for a blind experiment, end of story.

Stickley did, however, invite Zimmerman to visit DARPA's Virginia-based office to discuss the final experiment.

It was, Zimmerman later recalled, one of the odder professional meetings of his career. Stickley greeted him at the entrance to the building wearing a denim shirt with the DARPA "microtechnology office" insignia. For unknown reasons, the DARPA scientist declined to even invite Zimmerman inside the building. Instead, they sat at a table outside the building, next to a busy street, to go over the final experiment. The discussion quickly grew acrimonious.

Stickley informed Zimmerman that DARPA wouldn't allow a review committee to be involved in the experiment, and McDaniel wouldn't allow outside scientists to participate. Anyhow, it was obvious just by looking at the data that triggering had occurred, according to Stickley.

"How will you ever get the isomer to decay quickly enough to do any good?" Zimmerman asked.

"Once we found out the electrons were involved, it changed everything," Stickley retorted.

"How's that?" Zimmerman asked.

"That's very sensitive national security information. We aren't disseminating it," Stickley replied.

What would be the mechanism for such a chain reaction, Zimmerman asked?

"We don't want people to know," Stickley replied mysteriously. He went on to explain that while the triggering experiments would remain open, the rest of the work wouldn't be released to the public.

Actually, given the way the hafnium program had operated to date, that probably didn't mean anything other than business as usual. After all, McDaniel hadn't published any of his results, and Stickley still wasn't releasing the independent reports reviewing McDaniel's and Collins's work. National security was becoming an increasingly convenient cloak for poor scientific results.

That was the end of the conversation.

As responses slowly trickled into DARPA, the overwhelming consensus was that the final TRIP experiment, as described, was either insufficient to resolve the controversy over hafnium triggering, or simply an exercise in futility.

Livermore's response to DARPA's final experiment was simple rejection.

"Given that both sides of the argument have already

made a stand, there is a significant potential for conflicts of interest affecting the results," they wrote to Stickley. Rather, they argued that the Department of Energy should select an independent investigative team to conduct a counter-experiment.

They received no response.

The scientists from the Institute for Defense Analyses offered a more philosophical reply. They had at one point considered such a final experiment, but quickly realized there would always be some loophole. DARPA could not legislate scientific progress; it was the scientific method, and not a "proof of principle" demonstration, that would ultimately resolve the isomer-triggering debate, they noted.

They concluded with an eloquent and simple statement on the process of science.

"Neither we nor anyone else can predict in advance how nature will judge our efforts," they wrote. "Instead, we must rely on the usual trusted (but imperfect) criteria employed by the scientific community to judge emerging results: peer review and reproducibility across diverse research groups. Both of these are inherently retrospective, rather than pro-spective, activities."

The undated letter was signed simply: Jim Silk, Bohdan Balko, Dave Sparrow.

IN THE FINAL two weeks of September 2004, the United States faced a number of growing threats to national security. Iraq's reconstruction was supposed to be a priority, but rather than cranes and bulldozers, televisions around the world were filled with images of Western hostages begging for their lives, or in some cases the aftermath: gruesome images of beheadings. Iraq was rife with suicide bombings, and a new acronym, IED, or improvised explosive device, was entering the popular vernacular.

And if no one could find weapons of mass destruction

in Iraq, there was suddenly commotion over a different kind of weapon, with a flurry of activity in Washington that extended from the Pentagon, to the Department of Energy, to Congress, and finally to the national laboratories. DARPA had to find a laboratory with X-ray facilities willing to let Patrick McDaniel conduct his final experiment, and that was proving difficult. McDaniel had his eye on the Advanced Light Source at Lawrence Berkeley National Laboratory, a Department of Energy facility. Like with all experiments, McDaniel had to subject his proposal to peer review to get access to the facility.

Problems arose, perhaps inadvertently, when William Herrmannsfeldt contacted a couple of scientists at Berkeley to enquire about the state of McDaniel's request, which was being reviewed by an anonymous committee of scientists. One of the Berkeley scientists, in turn, asked if Herrmannsfeldt, and perhaps others involved in the hafnium controversy, would submit their own reviews of the McDaniel proposal, in addition to those scientists who had already been asked to look at the proposal. Herrmannsfeldt and several other scientists quickly agreed.

In the end, McDaniel's proposal was rejected: It was deemed unworthy of beamtime at the facility. The additional comments didn't even matter—the experiment had received resoundingly negative reviews, according to Berkeley.

That might have been the end of the story for the final experiment, except that shortly after McDaniel received the rejection, a Berkeley scientist sent out alarming news to the hafnium critics. A conversation had apparently taken place between Tony Tether, the head of DARPA, and Spencer Abraham, then the Secretary of Energy.

It was fortuitous timing in the world of fringe science: Tether was trying to save the hafnium bomb, and Abraham had approved a new review of cold fusion. There's no way of knowing for sure what precisely the two men discussed—or if the cold fusion review was brought up—but the hafnium experiment was definitely a topic of discussion.

When I asked DARPA Tether's spokeswoman if he had indeed spoken to Abraham about the hafnium review, I was told, "no such conversation took place." Perhaps the two men were simply good candidates for the CIA's psychic program, because one thing did happen according to a Berkeley scientist: Abraham's and Tether's offices contacted the lab with an unprecedented demand. They wanted the list of the anonymous reviewers that had panned McDaniel's proposed experiment. The Berkeley Advanced Light Source, as a Department of Energy facility, didn't have much choice other than to hand over the names.

The Tether-Abraham move was surely an "end-run" around the peer-reviewed rejection of the McDaniel experiment, the scientist at the Berkeley lab lamented in an e-mail to other government scientists. But that wasn't the end of the maneuvering. Around that same time, a Pentagon official from the Nuclear Matters Office called the Livermore lab out of the blue, expressing concerns about the military applications of hafnium-178. He also wanted to know about the status of the final experiment. In the meantime, on Capitol Hill, the defense committees' staff debated legislation that would direct the "Secretary of Defense and the National Nuclear Security Administration to initiate an independent technical review of the [Stimulated Isomer Energy Release] SIER program and other nuclear isomer research activities."

That review would look at all the isomer activities to date, and determine what to fund, and who should be funded to do the work. The proposed language acknowledged one important thing that scientists had been telling DARPA for the past several years. "The reportedly successful triggering experiment in 1998 has not been replicated."

14

BOOM OR BUST

THE CITY OF Albuquerque is a vivid reminder that man's power over the physical world, even in harnessing the energy of the sun, will forever pale next to the explosive forces of nature. Over 30 million years ago, the land under what constitutes present-day New Mexico began to split apart as fault lines deep under the earth moved in opposite directions, dropping the ground some 30,000 feet, carving out the Rio Grande, as the surrounding area rose up in volcanoes.

In 1945, the world's first nuclear device was exploded less than one hundred miles outside Albuquerque, releasing a force that was awesome to behold for a few split seconds in time, but whose remnants nearly half a century later are little more than bits of radioactive glass scattered across an otherwise desolate field of sand and rocks. That event—whose picture graces the office wall of many a weapons scientist—did not leave an indelible mark on the landscape. After over half a century, the effects of the world's most powerful weapon don't seem so great compared to the creation of the Rio Grande rift that extends from New Mexico to Colorado. It may have taken the earth's destructive power a bit more time and patience, but nature, as the saying goes, always bats last.

Located in a valley illuminated with brilliant shades of burnt orange and red rock, and framed on the eastern edge by the Sandia Mountains, Albuquerque today lies in one of the most beautiful locations in the American Southwest. Driving in at night, I once saw an almost biblical image of desert lightning striking the sunken city, but the morning in May 2004 that I flew in to see hafnium trigger, the air was crisp, the sky was clear, and every particle of sand and rock seemed to rise up from the desert to meet the descending plane.

After terse negotiations, DARPA had agreed to allow me to spend two hours with Patrick McDaniel at Sandia. I was finally going to see hafnium trigger, or not. Of course, observing a hafnium experiment was best described to me as somewhat akin to watching paint dry. There would be no boom or bang, just a few scientists sitting around watching data come in on a computer. But somehow I thought if I saw a hafnium experiment, it would clear away some lingering doubt. Besides, McDaniel's hafnium experiment could be one of the last ever conducted.

I also wanted to meet McDaniel in person.

When I had first called him to speak about hafnium, and asked him to describe his background, he gave a cryptic reply. He described himself only as a "a nuclear engineer and a Vietnam vet, who had spent time in the Air Force and was interested in national defense issues and how science plays into that."

It wasn't a lot to go on.

On the plane, I read through some notes, including a list I had pulled off the Internet called "The Crackpot Index." It was written by a well-known mathematical physicist named John Baez, who described it as "simple method for rating potentially revolutionary contributions to physics." A few of the entries sounded rather familiar:

- ❉ 20 points for suggesting that you deserve a Nobel Prize
- ❉ 20 points for bringing up (real or imagined) ridicule accorded to your past theories.
- ❉ 20 points for each use of the phrase "self-appointed defender of the orthodoxy."
- ❉ 40 points for claiming that the "scientific establishment" is engaged in a conspiracy to prevent your work from gaining its well-served fame, or suchlike.
- ❉ 40 points for comparing yourself to Galileo, suggesting that a modern-day Inquisition is hard at work on your case, and so on.

I glanced down at a recent e-mail that Carl Collins had sent to Martin Stickley at DARPA. "Production, I do not know, but recently saw results from Dubna where actual assay of a test production run yielded 17 percent concentration of the isomer," Collins wrote. "I can tell you that is big news, but news that will likely remain unknown while the 'high priests of innovation' snicker . . ."

In another recent letter, Collins wrote how government science has "been decided by a self-replicating group of 'elitists' in a secret 'Fraternal Lodge'" called the JASONs. "If they can miss the mark so far in their treatment of isomer triggering; then is this the first time or were earlier pronouncements of which they are so proud just as flawed?"

The taxi deposited me at a dusty drop-off point just outside the gate to Kirtland Air Force Base, where Sandia is located. As I waited for Neal Singer, the lab's public affairs officer who had helped negotiate my visit, I read *Thunder of Erebus*, the spy thriller based on Collins's isomer weapon. I needed to know how it all ended.

As best I could follow, the United States and the Soviet Union were about to square off over the only known reserve of the precious rubidium isomer. The superpower that controlled the world's rubidium supply would be able to build the fearsome isomer weapon and take control of the world's oil reserves. But even in Russia, officials disagreed on whether isomers were really worth all the fuss. Kasimir Bodin, a Russian navy officer, was arguing over the importance of rubidium with Yuri Timoshenko, the Soviet Union's minister of defense. It was silly, Bodin argued, to battle over a frozen wasteland for a theoretical weapon.

> "Meaningless, Yuri? This is insane. Joining the United States in battle on an ice continent to dig out some mysterious mineral? Insane."
>
> Timoshenko tossed the rest of his tea down and said, "Let me put it to you this way, Kasimir. Suppose, just suppose, there was only one source of uranium in he entire world, and that source was in Antarctica. Do you think we would fight to secure the sole source of uranium for ourselves?"
>
> Bodin fidgeted again. "I suppose so."
>
> "Then believe me, Kasimir. This rubidium isotope is ten times more valuable than uranium. Perhaps a hundred times. We cannot let it fall into American hands. . . ."

Just as the Soviet Union was plotting its final plans for world conquest using the rubidium bomb, Singer arrived at the scene in an unmarked white government minivan, ready to take me to the site of the hafnium triggering experiment. As we drove through town, we zoomed past boxy little offices for Northrop Grumman and Lockheed Martin; geometric squares that looked as if they had been airlifted straight out of northern Virginia and plopped down in the Rio Grande. Lockheed Martin, the largest U.S. defense company, operated Sandia under contract to the Department of Energy, and other defense companies have expanded in Albuquerque, picking up lucrative pieces of military work. Bell Laboratories had once operated Sandia for a mere dollar a year as a service to the country. Today, running Sandia, and the other national laboratories, is regarded as lucrative business that turns hefty profits.

As we drove, Singer began to talk in what was almost a nervous stream of consciousness that began with his wife's discovery that morning of his secret stash of bacon. They were Jewish, Singer explained, but he liked bacon, an affront to his more devout wife. But his wife discovered the illicit bacon while he was on his morning jog, during which he sprained his ankle. So he came back home limping with a sprained ankle, greeted by an angry wife and no bacon. Added to all that was hafnium. My trip out to Sandia had erupted in a protracted battle with DARPA over my access to the hafnium triggering experiment. That, too, surprised Singer, who couldn't understand why DARPA was so resistant to letting me come out in the first place. Science was full of controversies, he noted.

"It just seems that everyone is so emotional about this," Singer remarked, as we drove onto the Kirtland Air Force Base.

As we circled around, it quickly dawned on me that Singer, who was still talking nonstop, had no idea where we were going. McDaniel's work on hafnium, though touted as the next superbomb, was pretty obscure at Sandia. At one point, we walked into a modern building—an Air force training and simulation facility—thinking that the X-ray facility was somehow tucked inside or around back. Inscribed in granite above the entrance to the building was a quote by Vegetius, the fifth century Roman writer. "He who desires peace should prepare for war."

Barely glancing up from her work, a bored-looking secretary seemed confused when we asked about the X-ray facility. Wrong building. We got back in the car, and Singer began to look even more nervous as we circled around the base again, in search of the X-ray facility. "Relax," I told him. "I don't think it matters if I spend two hours with McDaniel, or just ninety minutes."

Singer shook his head.

"Mistakes aren't tolerated at Sandia," he said, half-jokingly, I presumed.

When we finally found the right building, it was hardly a surprise we missed it the first few times around. It was an older building that looked neither high-tech nor mysterious, just terribly dated—a feeling that was compounded as we entered a laboratory that struck me as eerily similar to a high school shop class. We walked past abandoned worktables, until we were greeted by Patrick McDaniel. Short and square, he politely stuck out a hand, burned deep brown by the New Mexico sun. He looked a lot like the legions of other military engineers I'd met over the years, in his late fifties, with a short-sleeved polyester button-down shirt, engineer's glasses, and close-cropped hair. Perhaps the only thing that seemed different about him was a sort of frenzied intensity.

It struck me that McDaniel was nearly alone in the lab, save a technician seated wordlessly behind a computer that operated the X-ray machine and an Air Force public affairs official. The man operating the X-ray looked up at me quizzically for a moment, but then turned back to his computer screen without uttering a word. Hill Roberts, McDaniel told me, was coming in later that day. McDaniel even looked a bit shy and uncomfortable at first as we walked toward the experimental set-up, but he quickly picked up the pace, running through an overview of his experiment in speech punctuated by intermittent "okays."

"This is very similar to a dental X-ray," McDaniel explained, pointing to the X-ray machine in the center of the lab. "Basically, it's a radiography machine, okay."

The X-ray machine was the size of a small shed, with a

small door at the front that provided access to the central chamber. It was miniscule compared to the massive synchrotrons at Argonne or in Japan. It was primarily used by the Air Force to test how microelectronics would be affected by radiation in space, McDaniel explained. He had convinced the brass to let him borrow it to trigger hafnium. It wasn't a glamorous machine, but it did look more scientific than a dental X-ray device.

McDaniel still defended that early experiment, however.

"I thought when I started, 'a dental X-ray, that's not a real irradiator,'" he said. "Okay, then I came over here and started doing the analysis, and it was hard to get back to where Carl was. You think that's easy to do, okay, but it's not."

So a dental X-ray is actually good? I asked.

"A dental X-ray machine is a pretty powerful X-ray machine while it's on," he said. "While it's on, the flux is very intense. Okay, Carl was very clever to see that. I would not have thought to look there. I've got a chart I can show you that compares all the accelerators at Sandia, okay. This gets more flux than any of the other places. What matters is flux over time. And because this runs continuously and getting the energies, this is the best place to do experiments," he insisted.

McDaniel's assertion was convenient, in fact, since no independent laboratory had agreed, thus far, to allow him access to facilities to conduct his final experiment for DARPA.

It occurred to me, watching McDaniel by himself at the controls of the X-ray machine, that there was a certain element of desperation. He was a government engineer afforded little opportunity to do original research. He had once described a PhD to me as nothing more than a "union card" (and presumably a credential he didn't have). The dental X-ray experiment was not simply a blip in an otherwise unremarkable career in science, rather it was a blip of science in an otherwise unremarkable career as a government bureaucrat. McDaniel looked at the X-ray machine.

"I never thought I'd have anything in *Physical Review* in my whole life," he said wistfully. "Okay, that's a big deal. It's the most prestigious physics journal in the world."

The X-ray machine was off, so McDaniel opened the door and led me inside, pointing to an aluminum disk screwed in place with four bolts. Behind the thin protective sheet of aluminum was the sample of hafnium-178, radiating away. Much better, at least, than a Styrofoam cup, I thought.

I couldn't help but pause to think about those precious few atoms of hafnium, which even over the short two hours I would spend in the lab, were disappearing. In thirty-one years, the world's supply of hafnium would be halved, and then halved again, and again, until if no one produced anymore hafnium, it would eventually just disappear off the face of the earth, never to be seen again.

Now, he was in the middle of a "beam off" run, counting the background radiation while the X-ray was powered down. Later, he would turn the beam on to count photons that would be evidence of triggering. Of course, over that past year, McDaniel had conducted a series of experiments at a much higher-tech facility at Louisiana State University, and independent observers monitored his work there. What did the observers question him about, I asked?

"They didn't ask any questions, they told me what I was doing wrong," McDaniel said. "Okay, that's really distracting."

McDaniel, I learned, didn't like to be distracted, and more importantly, he didn't like to be questioned. He paused for a second, squinted as if some thought had struck him, and then continued.

"Okay, one guy said, and I won't give his name, 'If the physics occur the way we were predicting, we'd have to take

Einstein's Nobel Prize back.' I said, 'Okay. I came back here and found three books that said this is what people predicted, and you should take Einstein's prize away.'"

He paused and squinted again, before continuing. The world has changed since 1975, McDaniel asserted, and science had come to view atomic physics "near the edges of atoms" in a whole different way. "Basically, what we observe is very significant wiggles in what were thought to be fairly flat cross-sections," McDaniel said, citing no particular scientific reference. "Those wiggles look a great deal like triggering cross-sections we see in triggering experiments."

Older scientists just don't understand this new physics, according to McDaniel. "There's a lot of people around who make decisions and have not learned that physics."

It seemed to me that McDaniel, in my brief time at his lab, had just told me that Einstein was wrong, most physicists were wrong, and hafnium triggering was perfectly logical. That sounded impressive, I thought. And though he hadn't yet used the word "hidebound reactionary" to describe physicists, his lecture still sounded pretty high up on the crackpot index.

I circled around to the X-ray machine and stared at it for a moment.

"How much triggering would you need to convince people?" I asked.

"Ten percent would be overwhelmingly convincing," McDaniel replied.

"But you had 15 percent," I pointed out. That's, after all, what Martin Stickley had told everyone. Why wasn't that convincing?

McDaniel hesitated and sighed. "Yeah, right, it would be convincing if we could repeat it, but we think there was a screw-up in the detectors. That's never happened before. We've got new detectors," he said, pointing to the ones surrounding the hafnium sample.

"Why was it a screw-up?" I asked.

"Because it went away, we couldn't repeat it immediately. It wasn't repeatable. It was 30 percent sometimes, but the average was 15 percent."

If the hallmark of a scientific result is reproducibility, and McDaniel's results couldn't be reproduced, then what he had was suddenly not 15 percent, but nothing. Zero. And if all the claims of McDaniel's "confirmation" of triggering were nothing more than an experimental artifact, then it suddenly wasn't clear if anyone had confirmed Collins's results.

We both stood in silence next to the decaying hafnium.

There was another issue I wanted to bring up with McDaniel; the issue that William Herrmannsfeldt had raised. Even if Collins's original results were correct, there was still the question of whether anyone could get more energy out of hafnium than was being used to trigger it. If only one out of every few hundred photons was triggering the hafnium, it wasn't likely to be a very effective bomb. This, in the most basic language, was the question of "breakeven."

"Can you do that?" I asked McDaniel. "Can you achieve breakeven?"

McDaniel suddenly grew quiet and glared silently, ill at ease with the question. Clearing my throat, I asked it again. For a full minute, we both stood there in silence, next to the hafnium, the big X-ray machine, and the two public affairs officials, who shifted uncomfortably in their seats. I thought again about those precious atoms of hafnium, which were decaying away even as we sat there in silence.

"I don't want to answer that question," McDaniel finally snapped abruptly, turning back to the X-ray machine.

I was a bit dumbfounded. After all, wasn't that the whole point, the whole reason everyone was interested in hafnium? If you can't get more energy out than you put in, then you can't build a bomb, or much of anything else for that matter.

I stopped to glance back at the X-ray machine that held the microscopic hafnium sample. I hardly felt like I was in the middle of a Manhattan Project.

"Well, why don't you want to answer the question?" I asked, not quite knowing where to go from there, and hoping to at least further the conversation a bit. After all, I couldn't think of anything else to ask, and I still had over half an hour left of my allotted two hours.

"I won't answer that!" McDaniel retorted. He paused in silence for a moment before beginning to speak again. "There's a reason I don't want to answer that, it tells too many other things," he added with a conspiratorial tone.

Other things? The only thing I could think of was the other reason why hafnium wasn't going to be a bomb. Even if you could get more energy out of it than you put in, you'd have to find a way to get a chain reaction, like with nuclear fission. McDaniel had once hinted to me in an e-mail that there were clandestine proposals to make such a chain reaction, but no one had presented any clear ideas for how that would work. When McDaniel had suggested one mechanism for a chain reaction to the JASONs back in 1999, the scientists had mused that his idea would require nothing short of a miracle.

Clearly the conversation wasn't going anywhere, and if he wasn't going to answer any important questions, I decided I might as well ask McDaniel about the one subject I knew would set him off: the nuclear hand grenade. After the *Washington Post Magazine* had run a picture of the nuclear hand grenade, McDaniel had told me the illustration, which was taken from Martin Stickley's briefing at the first hafnium production panel meeting, was all a mistake. He would never say why.

"So, what was wrong with the picture of the hand grenade?" I asked, sitting down on a stool by an empty worktable.

McDaniel's face contorted and turned bright red. "That picture made the whole subject inflammatory," he sputtered.

"If that picture wasn't there . . ." His voice trailed off for a moment, before beginning again. "It was a closed meeting, and he didn't have the slides reviewed! Martin made a mistake! The two-kiloton grenade has produced a lot more fire than has been useful for anyone. It was a liability. Martin made a mistake."

I got up from the chair and took a step back, realizing that the conversation was beyond salvation. McDaniel was clearly agitated, but I didn't care. I was the one who had to catch a three-segment redeye flight back to Washington that evening. The Sandia and Air Force press officials looked nervous and shifted in their seats. McDaniel continued to rant about isomers, and then stopped midsentence and paused, staring at me.

"Did you know that cold fusion could solve our radioactive waste problem!" he shouted, his voice rising to a crescendo.

Air swooshed down from the air-conditioning, but I just stood and stared, blinking and wordless. The technician manning the X-ray machine even looked up to see what was going on. McDaniel added something about permutations of radioisotopes in cold fusion experiments, and took a step toward me. I took a stumbling step back as I tried to process this complete non sequitur. I couldn't quite figure out how cold fusion has jumped into the conversation.

McDaniel kept talking, and suddenly, it was all clear to me. The crackpot index was wrong, because it assumed crackpots were crazy. Carl Collins wasn't a crackpot. And McDaniel wasn't at all crazy, he was simply a member of the Pentagon's scientific underworld: a group of government officials who sincerely believed that a cabal of elder physicists had a stranglehold on modern science. The members of the Pentagon's scientific underworld were on a holy crusade to save the United States from the threat of foreign armies advised by psychics and equipped with isomer bombs and teleportation devices. It was a world of fantastic imaginary weapons, ranging from mobile vans in Iraq that were producing chemical weapons to secret hafnium facilities buried deep in Russia.

It was a scary world, and I suddenly wanted to leave it.

Singer and the other Air Force public affairs official quickly stepped in and announced that the interview was over. That was fine with me, since it was clear there was nothing left to say. Singer offered to take me back to the airport, which I thought was a generous offer to the alternative of being dumped off at the pickup point beyond the gate.

As we drove back, Singer talked again, nonstop. I couldn't quite remember what he said, except that he, too, appeared a bit shaken. I concentrated on the scenery; from the plane Albuquerque appeared impressive and imposing, but from the ground it looked like a strip mall dropped in the desert.

"I don't think he liked you very much," Singer said, as we drove into the airport lot.

I shrugged my shoulders. "I'm sure it's nothing personal," I replied, exiting the car.

When I got back to the Albuquerque airport, I checked my e-mail. There was a note from Tony Tether, DARPA's director. He said he hoped that I had "enjoyed" my trip to Sandia. "However I wanted to make sure you understood that no further visits will be authorized."

I couldn't imagine wanting to visit another hafnium experiment.

IN OCTOBER 2004, DARPA was forced through a Freedom of Information Act request I filed to release a copy of the famous Stone report that Martin Stickley had commissioned to review Collins's experiments. The report, which Stickley had subsequently refused to acknowledge or release to the critics, was emphatic in its conclusions. In reviewing paper after paper, the Stones found only inadequate statistics, poor procedures, and inconsistent data. Again, and again, they rejected the alleged proof of triggering as nothing more than bad data.

In the meantime, with the DARPA hafnium program coming to a close, its participants began to turn in their final results. In the fall of 2004, James Carroll submitted his report, which wrapped up a series of different hafnium triggering experiments that ranged from Canada to Japan.

"At this point, no support was found for claims of triggering of (178/sq cm)Hf near 10-keV," Carroll wrote in his final report to DARPA, concluding his eighteen-month work on the Stimulated Isomer Energy Release program. In over four experiments, Carroll never found even the slightest evidence that hafnium triggered.

Even Hill Roberts, Collins's erstwhile colleague who once showed pictures of hafnium's divine perfection, was forced to come to similar conclusions.

"The data from the March 2001 burn-up experiment has now been fully analyzed. No burn-up due to synchrotron radiation was detected in this experiment due to both experimental limitations and the unlikelihood of ever observing burn-up in the Hf-178m2 system for triggering enhancements of less than 10 percent. We suggest looking at alterative isomer systems with shorter half-lives, such as the Lu-I77m isomer."

And while McDaniel promised several times to write up the results of his experiments in a formal Sandia report, no one ever saw it. But a few months later, I obtained a copy of a second report by the Stones, this one a review of McDaniel's work at Louisiana State University. McDaniel had claimed to the Stones that his two experiments at Louisiana were part of "a very successful overall program in that they replicated the work of Dr. Carl Collins at the SPring-8 synchrotron and improved the estimate for the triggering cross section near the L3 edge in hafnium by a factor of two."

What the Stones found, on the other hand, was a simple and embarrassing error. A drift in McDaniel's detector equipment corresponded with his alleged finding of triggering. "Our analysis clearly does not support these results, or the conclusion that they replicate the Dallas group experiments in any way," the Stones wrote. "The Sandia report is extremely brief and gives few details of their data treatment."

The report continued: "After detailed analysis we can find no evidence from any part of the data taken during the November experiments at the CAMD Facility to support the claim of enhanced decay of Hf-178m2 triggered by the presence of X-rays in the energy range 9509 to 9595 eV. That the X-ray flux available at CAMD is lower than at other facilities may be a cause of this negative result. However the lack of strong evidence of triggering, proportional to X-ray flux, from this and other experiments, remains a basic problem for proponents of the phenomenon."

There was still one final hafnium gap, so to speak. The 1998 dental X-ray results had been widely debunked in the scientific literature, and hafnium triggering reduced to a laughingstock among those in the know. But for the outside world, and for the press, the question of whether hafnium could trigger still fell into a trap. Some said it worked, others said it didn't. Journalism, which relies on "balance," has never dealt well with science, which relies on consensus. And the scientific consensus had clearly rejected the hafnium results. Worse, most scientists in their publications were addressing—in the most incomprehensible language—the very narrow scientific question of whether hafnium-178 could be triggered by low-energy photons. Why couldn't scientists just say that it couldn't be a bomb? That gap was about to be filled, not by an American scientist, but by a physicist in Russia, the country that presumably was "ahead" of the United States in the imaginary isomer arms race, at least according to DARPA.

In August 2005, Eugene Tkalya, a Russian physicist, published an unusual scholarly article: a scientific attack on the "hafnium bomb." To that point, the papers criticizing Collins's work in scientific journals had been very narrowly focused on the purported results of the "dental X-ray experiment." The scientific journals don't typically address issues of potential applications. Tkalya apparently had tired of governments in both countries worrying over an imaginary bomb and set out, at least from the scientific side, to do something about it.

"Recently, there have been reports in the mass media about plans to build what became known as an 'isomeric bomb' based on 178Hf," Tklaya wrote. "What all the publications are speaking about is no less than the possibility of building a radically new weapon that does not fall under a single article of the existing nonproliferation treaties."

Tkalya's article went on to analyze, in excruciating scientific detail, the implicit assumption in Collins's work, to demonstrate where basic physics was being violated. It was all just bad science, he concluded. "Summarizing the results in the present report, I would like to note the following. Theoretical calculations and the analysis of the existing experimental data suggest that the hafnium problem, as presented by the works of Collins's group, does not exist," Tkalya wrote. "The hullabaloo over the hafnium bomb was due to meaningless experimental data and the incompetence of certain individuals rather than to the real possibility of building a radically new weapon based on 178Hf."

Congress, at least at that point, seemed to agree. Motivated by the publicity over the hafnium bomb—and the picture of a nuclear hand grenade—Congress stepped in. And in the fall of 2004, the House and Senate finished up work on their separate versions of the defense bill. While the House and Senate quite typically disagree on many issues, that year both arrived at one firm conclusion. The hafnium bomb had to go.

"The committee agrees with the majority of expert technical opinions that this effort, though carrying a large potential payoff, should be a smaller fundamental research effort at this time," the Senate bill concluded. No funding would be authorized to continue work on the hafnium bomb.

The House version of the bill, took an even more extreme stand.

"Given the significant policy issues associated with any eventual use of an isomer weapon and given the inability of distinguished scientists to replicate the reported successful triggering experiment of 1998, the committee believes that the Department of Defense should not be engaged in this research," the House authorizers wrote in their version of the defense bill. "The proper agency to investigate the feasibility of this technology is the National Nuclear Security Administration and its national laboratory complex. The committee questions the utility of this research in any circumstances and is particularly skeptical of research into nuclear isomer production before triggering is shown to be possible."

The final bill, passed in late November 2004, ordered DARPA to terminate the hafnium program.

Hafnium was dead.

From Romania, to Texas, to the Pentagon, hafnium had finally been defeated.

Or so it seemed.

JUST ONE WEEK after Congress killed the hafnium bomb, Ehsan Khan sent out his final letter to the members of the hafnium production panel. In it, he remarkably announced that the panel's work had concluded that hafnium could be produced for as little as $3 million a gram. "There appears to be only one way, at present, to enrich sufficient quantities of feed material from which the isomer is produced," Khan wrote in the letter, which was also signed by Paul Robinson, the head of the Sandia lab. "This is the Plasma Separation

Process (PST) that has been traditionally used for isotope separation."

The Department of Energy had originally developed the plasma separation process to enrich uranium-235, but its funding was cut off for production in the late 1980s, Khan explained. A Georgia-based company named Theragenics then leased the production facilities, investing $28 million.

"Theragenics currently operates the PSP at its isotope production facility in Oak Ridge, Tennessee," Khan noted.

It was hard to understand the purpose of Khan's lengthy letter about an obscure company named Theragenics. After all, as Khan also noted, Congress had cut off funding for the hafnium program at the end of the fiscal year. What difference did it make?

That explanation soon became clear.

In November 2004, Congress also passed the annual energy bill, which buried deep in legislative language, contained the following passage: "From within available funds, $5,000,000 for National Energy Technology Laboratory to use the Plasma Separation Process to develop high energy isomers and isotopes for energy storage and utilization."

What company was to receive the $5 million in congressional largesse? Theragenics.

ON HEARING OF congressional action to kill the hafnium bomb, Collins, who had traveled so far with his dental X-ray machine, wasn't quite ready to give up.

"Very soon will be circulated within the 'corridors of power' an analysis conducted of the Argonne failures by eminently respectable individuals showing that what was marketed as null was actually a 4.5% triggering—the largest seen by anybody," he wrote to me. Plans were already underway for a "micro-detonation," to prove that a chain reaction was possible as well, Collins declared.

In a series of incriminating e-mails, Collins accused me

of helping to engineer the death of hafnium. I had become yet another member of a vast conspiracy to thwart his efforts. The final sentence of the very last e-mail that Carl Collins would write to me was written in bold and highlighted with italics: ***"Remember, when triggering hafnium is outlawed, only outlaws will trigger hafnium."***

I thought that was an appropriate way to end our correspondence.

READING THROUGH MY large collection of hafnium files at home in Washington, I thought back to my last hour in Collins's dusty lab down in Texas. Sitting with the would-be father of the isomer bomb in his office, I found myself gazing at an oil painting hanging on the wall, its frame coming off at one end. Incoherent streaks of browns and oranges emanated out from the center of the painting.

"It's an eclipse," Collins explained, when I asked him about the artwork. After two solid days of interviews, I was getting ready to leave his office and head back to Washington. But the painting had caught my eye.

It was, as I correctly guessed, something Collins himself had painted.

Collins explained that his approach to composition was taken from his father, an amateur photographer who won prizes in local competitions. Collins, however, favored impressionism and painting.

"The idea," Collins said pointing at the picture, "was there's inspiration or motivation in science that suggests impressionist painting or painting style."

Impressionist painters, according to standard encyclopedic descriptions, often took science as their subject, using broad strokes in place of detail to convey a "subjective" image. Looking at the painting, I could perhaps see Collins's eclipse, or something entirely different.

There was another reason Collins likes oil paintings.

They are "infinitely correctable—you can change it," he said with a slight smile.

Collins took the same approach to his science. With each criticism, he would revise his arguments just a bit. Argonne failed to get triggering, he said they looked at the wrong energy levels. They changed the energy level; he said they looked at the wrong emission lines. And when all that failed, Argonne did get triggering, and just didn't know it. It was the same with hafnium. It was an energy source. It was a gamma-ray laser. It was a bomb. It was an eclipse. Hafnium could be whatever it was needed to be, forever metamorphosizing to fit the needs of the day.

As we stepped outside his office, I turned to a data graph taped to the wall in the hallway near the entrance. Collins pointed to the graph, which showed data from one of his more recent experiments. Data points dotted the graph where Collins claimed evidence of triggering, and he noted the famous "130-KeV" line that Argonne couldn't see. "Is there something there or not?" Collins asked as I moved in front of the graph, taking a long, hard look. "If you looked at this, would you say this is just noise?"

Collins paused and looked at the graph himself and shrugged his shoulders. "Well, people have different opinions," he said.

Looking at the data points, in fact, one could either draw a line connecting the points in spikes, as Collins did in his articles and on the graph taped to the wall, or without that line, it might just appear as random dots on a graph. Nothing more than "fly droppings," as Donald Gemmell had once derisively called them.

"Sort of squint at that picture and look at it at an angle," Collin instructed. He paused as I stared intently at the graph. "Do you see something there or not?" he asked again, bending in to look at the graph himself.

"Well, some people do and some people don't," Collins said after a moment.

Staring at the graph again, I squinted hard and tried to look and see the line that Carl Collins, Patrick McDaniel, Martin Stickley, and Nancy Ries all saw—that line that would fulfill their dreams of powerful bombs.

Perhaps physicist Richard Feynman best explained what eventually happens to those dreamers. "We've learned from experience that the truth will out. Other experimenters will repeat your experiment and find out whether you were right or wrong. Nature's phenomena will agree or they'll disagree with your theory. And, although you may gain some temporary fame and excitement, you will not gain a good reputation as scientist if you haven't tried to be very careful in this kind of work."

It was late and I needed to get to the airport. I turned to thank Collins for his time, and took one last look at his graph.

Could I see that line?

I looked and looked, but in the end, all I saw were dots.

A Hafnium Ending

IT MUST HAVE been a stunning sight. Reflected off a guard's flashlight, the shiny metal cast back an eerie silvery glow. It was Saturday in the sleepy town of Russe, Bulgaria, when the border patrols stopped a car holding three Romanians and a Bulgarian driver. The group was preparing to cross the Friendship Bridge, which links two countries whose relations have never really been that friendly.

A routine check of the car turned up an unusual find: over three kilograms of pure hafnium.

What did the borders guards think as they searched the driver and spotted the hafnium metal tucked under his jacket? Perhaps they had no idea what the misshapen lump of mystery metal was. Did they instinctively reach out to touch the metallic object, or did they hesitate, worried that it might be some dangerous, radioactive substance? How did they even know to search the driver? Did someone tip them off?

The Bulgarians seemed almost in awe of the precious metal—slick and shiny even in its unrefined form—and anxious to advertise to the world their diligent checkpoints.

"Bulgarian police have arrested four men trying to smuggle into Romania a substance scientists say can be used to make powerful explosives or 'dirty bombs,'" Reuters reported on September 19, 2005.

"It is a very rare element—very beautiful and extremely expensive," Marina Nizamska, of Bulgaria's Nuclear

Regulatory Agency, later told Agence France-Presse. "It can be used as a material for rockets and even for a bomb," she said, according to the Reuters report.

The hafnium capture turned out to be good publicity for the upwardly mobile Balkan country, whose semiporous borders have often proved a convenient transit point for illegal goods. "Bulgaria thwarts smuggling of nuclear material," announced one headline, as reports of the capture spread over the international wires. The Bulgarian authorities even went as far as to call a special press conference in tiny Russe to explain the "great importance" of the intercept for Bulgaria's national security. Like the Czech Republic's desperate attempts to impress the West with its reports—never verified—of 9/11 mastermind Muhammad Atta's visit to Prague, there was a ring of self-promotion in the Bulgarian government's jubilant announcement of captured hafnium. What was even more concerning, perhaps, were the subsequent claims made in various news reports that the hafnium was destined for the Middle East.

Probably no one involved in the affair quite got the irony that the hafnium was traveling to Romania, the country where, some might argue, the isomer bomb was first born. But in the spirit of grandiose claims, one of Scotland's leading newspapers, the *Sunday Herald*, did run a full article asserting that the hafnium capture was tied to Iran's nuclear weapons program:

> The *Sunday Herald* has learned from Romanian sources that an Arab-dominated Bucharest mafia was the intermediary in the hafnium deal. The sources could not give the intended final destination of the consignment, but the "working hypothesis" of Balkan police forces is that "it is linked to Iran's nuclear quest." Then again, there are always al-Qaeda armourers keen to buy dirty bomb material.

Shortly after the seizure, a Romanian journalist contacted Peter Zimmerman, by then a professor at King's College in London, to ask for his thoughts on the recent events. Visions of hafnium bombs must have clouded the journalist's head.

Since the hafnium seized was regular ground-state hafnium—the silvery-gray nonradioactive metal—and not the hafnium isomer, Zimmerman pointed out that it likely posed no military threat, at least not as a weapon. Even if it were hafnium isomer, it couldn't be used as a bomb, he noted, going on to reprise briefly the scientific debate over isomer triggering. As for what threat, if any, regular old hafnium might pose, Zimmerman suggested one possible danger. "If dropped from a couple of meters onto your foot, 3 kg of hafnium could cause injuries."

The hafnium seizure took place in September 2005, as I was finishing up the manuscript for this book; the same month that the United Nation's nuclear body, under pressure from the United States, declared Iran to be noncompliant with its nonproliferation obligations.

What difference do some half-baked assertions about an Iranian hafnium bomb make? Of course, few will forget the famous "sixteen words" that haunted the administration long after it gave up looking for Iraq's nonexistent weapons of mass destruction. "The British Government has learned that Saddam Hussein recently sought significant quantities of uranium from Africa," Bush told the nation in his 2003 State of the Union address.

That information, as we now know, was based on a document so ludicrously phony that triggering hafnium with dental X-rays looks almost reasonable. Common sense should dictate that reports of captured hafnium destined for Iran would never make it as far as a presidential speech, but is a badly forged document on yellowcake any better?

Hafnium, if nothing else, has taught me that when it

comes to imaginary weapons, there are no limits to silliness, and a story that should be comical has more elements of tragedy than pure farce. Zimmerman, more than once, joked to me about how the highest level of classification is TS/E, or Top Secret/Embarrassing.

Is it all a joke?

Hafnium is a funny story about funny people who believe in funny things, and if weren't for the fact that the main part of the story, between 2002 and 2005, took place against the backdrop of the war in Iraq, I would say it's an *extremely* funny story, with some moderately important lessons about the dangers of politicized science. But, the hafnium bomb is also a story about the depths of self-deception and the willingness of government officials to believe in threats that don't exist. In Iraq, the pursuit of imaginary weapons is anything but funny.

In the months leading up to the 2003 invasion of Iraq, President Bush again and again provided the following logic to the American public. While the United States didn't know for sure that Iraq possessed weapons of mass destruction—and in fact many experts vehemently objected to the "evidence" of nuclear and biological facilities—the threat of inaction was too high. "Facing clear evidence of peril, we cannot wait for the final proof—the smoking gun—that could come in the form of a mushroom cloud," he argued.

Scientific facts really are important. As we now know, Bush and his closest advisers probably could have declared a toilet seat a critical centrifuge. And perhaps the same goes for the hafnium believers. Collins, as one cynical scientist suggested, probably could have looked at a bank statement and seen it as evidence of triggering.

What's a few million dollars compared to the fate of the nation? Of course, the fallacy of that logic is that if you aren't honest about the odds, the deck is stacked against you. We now know that Iraq's weapons of mass destruction were more imagined than real and that the hafnium bomb is little more than a dream.

FOR TWO YEARS I had followed the hafnium controversy, from its apex as a Pentagon program gearing up to spend in the billions of dollars, to its near extinction at the hands of Congress. But I hadn't yet learned the most important lesson of the scientific underworld: it never really goes away. Shortly after the hafnium capture at the Bulgarian border, a cold fusion scientist forwarded me a copy of an e-mail sent out by James Corey, the Sandia intelligence official worried about isomer bombs back in the 1990s. The e-mail was a proposal to have the isomer believers join forces with cold fusion scientists.

"I will see if I can't get a mailing list for isomer researchers and invite them to join," wrote Corey, who had just founded his own cold-fusion company called Research Institute of the Ozarks, LLC.

Other than the rainbow coalition aspect of fringe science, I couldn't understand what hafnium and cold fusion had in common. Cold fusionists were convinced they could save the world from spiraling into an energy quagmire; the hafnium mafia pretty much wanted to be able to blow up the world. My mind momentarily logged flickering images of meetings held in exotic locales, with members of the grand hafnium-isomer coalition competing for attention—the hafnium gang drawing cartoon diagrams of nuclear hand grenades to the horror of their earth-friendly cold-fusion counterparts. It seemed like an unholy alliance if there ever was one.

There was no word, however, on whether anyone had joined up yet.

As for cold fusion, in November 2004, the Department of Energy released the results of its second review of the field. Eighteen reviewers looked at progress made in fifteen years of cold-fusion experiments carried out by a dedicated band of scientists. Some of the reviewers thought there was something to it, some didn't. None of them could really agree. What had all that work produced? "Fifteen years have gone by and according to one estimate more than $60 million has been spent on cold fusion research around the world," one critical reviewer noted. "There is a little to show for this."

All that time, all that money, and what did they have to show?

In the course of this book, hafnium is often compared to cold fusion—both as an example of a pathological science, and other times, simply as a metaphor for bad science. I certainly don't claim to know whether cold fusion might someday pan out, but the parallels with hafnium are clearly there: magnificent claims of groundbreaking results that challenge conventional physics; difficulties experienced by independent researchers in reproducing those results; and finally, the inability to identify scientific theories that could convincingly explain those claims. In that respect, hafnium and cold fusion are inexorably linked, for better or worse. Another, and equally important, element that links the two is the willingness of a group of government officials to entertain these ideas and even promote and support elements of the scientific underworld.

Why do they do it?

Perhaps in the final analysis it has something to do with being among "fellow travelers." Like the communist sympathizers who never joined the party, cold fusion, hafnium, and other areas of "fringe science" attract government officials and scientists who never join the movement, but remain in sympathy to it. The officials who believed in hafnium were often the same ones who believed in other quixotic endeavors. There isn't one great division between mainstream and fringe science, but rather, degrees of variation that blur, and sometimes overlap with one another. But the starting point for all of this, and the gateway into the scientific underworld, is a certain willingness to believe the impossible, and to be sympathetic to those who try. In moderation, that's not a bad thing, but in the extreme, it's a recipe for disaster.

This book should not be read as a clarion call to rally forces against fringe science, but rather, a warning that without a government willing and able to turn to scientific advisors, there is little chance for sound public policy, particularly in the realms of national security. I am not at all bothered by Hal Puthoff pursuing zero-point energy at his lab in Austin, or by legions of psychics attempting to locate foreign threats (and if they could help spot a few IEDs, all the better). And someday, perhaps some scientist from the underworld will really stumble onto a breakthrough. I wish them luck. The world may benefit from Hal Puthoff, but it also desperately needs Peter Zimmerman, and unless government is willing to listen to scientific advisors, it faces a future filled with imaginary weapons.

I spent two years in the Pentagon's scientific underworld, and in the end, I can't say with certainly that an isomer bomb won't ever exist, or that cold fusion will never be proved real. For that matter, I don't really know whether psychic plants can communicate the location of Soviet submarines, though I have my doubts. I will keep an open mind, but the pursuit of a goal, however noble, or the fear of a possible weapon, however frightening, should not lead government into self-delusion. Maybe we don't know with certainty that an isomer bomb will never be built, but we do know that Collins's experiments—as reviewed by the experts—were certainly flawed.

But suppose isomer triggering eventually turns out to be possible? Would the government have been wrong not to take heed of Collins's work as flawed, but prescient? That answer is no. I am reminded of what Irving Langmuir said of pathological science: "All I know is that there was nothing

salvaged at the end, and therefore none of it was ever right.
. . . You can't have a thing halfway right."

DOESN'T THE HAFNIUM story end well, I've been asked? After
all, didn't the process work and good science triumph over
bad? Spurred by the media publicity over a scientific and
political fiasco, Congress stepped in and ended the DARPA
isomer bomb, so isn't there a hafnium ending? Is it time to
write an epitaph to the hafnium bomb?

Of course not.

Even as the hafnium dreams receded to the background,
its supporters and fellow travelers were still out there, and
perhaps for that reason alone, hafnium—and the rest of
the scientific underworld—is here to stay for a long time
to come.

By the fall of 2005, the danger of a multibillion-dollar
hafnium program seemed to have diminished and among
most scientists, at least, hafnium had become something of a
bad joke. But has the hafnium bomb really disappeared into
the trove of the Pentagon's imaginary weapons, ranging from
the CIA psychics to the Air Force's studies of teleportation?
Well, not exactly.

As for the hafnium budget, DARPA, despite its last-minute appeals, was forced by Congress to give up its pet project.
Sort of. In a final plot twist to the never-ending hafnium saga,
the mysterious $5 million was slipped into the Department
of Energy's fiscal year 2005 budget earmarked for isomer
production. It was never quite clear exactly what the money
was intended for, but the suggestion was that it was being
used to keep the hafnium bomb alive.

Where are the crusaders for the hafnium bomb now?

In late 2005, Tony Tether was entering his fifth year as
the head of DARPA, making him one of the longest-serving
directors in the agency's history. DARPA has not done badly
in some areas. It held its second robot race across the desert

in the summer of 2005, and this time, several driverless cars motored triumphantly across the finish line. DARPA hailed the race as more proof of its "DARPA hard" motto.

Then again, no one ever suggested an unmanned Hummer violated the laws of physics.

If Tether still supported hafnium, he wasn't talking about it publicly, but he still seemed to be smarting from the congressional rebuke, because when I last ran into him at a hearing on Capitol Hill, his face went pale, his eyes opened wide, he turned heel and quickly scampered away.

Martin Stickley, hafnium's chief patron at DARPA, was still at the agency, and according to those who knew him, still trying to save hafnium. He never did agree to speak with me. At least officially he had moved on to other projects, like ultrabeam, a high-risk laser promoted, ironically, by a Collins nemesis named Charles Rhodes. Physicists generally regarded Rhodes, a professor at the University of Illinois in Chicago, as a far superior scientist to Collins, though many found his laser schemes almost as dubious as an isomer weapon.

I caught sight of Stickley once at Disneyland during DARPA's annual convention, but I decided that introducing myself wasn't likely to lead to a productive conversation. His motives for promoting the hafnium bomb, and his blind faith in Carl Collins, will always remain a mystery to me (though perhaps not to many of his colleagues, who had long since written him off as a scientific lightweight). Why was Stickley, in the terminology of physicist Richard Garwin, "hafniumized" by an imaginary weapon? I can only guess that Stickley found in the Pentagon's scientific underworld some hope that mainstream science couldn't provide, like the opportunity to be a part of something important, a world-changing discovery.

In either case, I was told that he has recently set his sights on cold fusion.

Ronald Sega, the former astronaut who was nominally Tether's boss, had received a big promotion, becoming in

2005 the undersecretary of the Air Force and the Pentagon's top official in charge of multibillion-dollar space satellite programs. That must have been good news for Sega, who no longer had to drag the talking phrasealator around to congressional hearings. He still smiled a lot.

At one point, while rummaging through the archives of the American Institute of Physics in College Park, Maryland, I even ran into Bert Schwarzschild, the reporter for *Physics Today* who had experienced Collins's wrath up close. Schwarzschild said he had recently overheard at a conference Stickley's partner-in-crime, Ehsan Khan, discussing the latest hafnium experiment plans. Khan was waxing enthusiastic, and promising good results. "These guys never give up," Schwarzschild remarked.

He was right.

I sometimes got news that the Khan and Stickley duo were still both making the rounds as part of the hafnium road show, often accompanied by the mysterious contractor, Tom Ward. In one meeting described to me by a cold-fusion supporter, the hafnium believers met at the Livermore lab in California to make another pitch to convince famed Star Warrior Lowell Wood; Mort Weiss; and John Nuckles, the former laboratory director, of the wonders of hafnium. Ward still claimed to have a promising theory to explain hafnium triggering, though he had never published his idea, or even tried to publish it, to the best of anyone's knowledge.

Khan and Ward were also handing out papers on cold fusion, which hardly surprised anyone, given how over-the-edge things had already gotten with hafnium. I never heard from Ward, who I e-mailed once requesting an interview, but Khan, when I contacted him, expressed initial enthusiasm about speaking with me. However, he prefaced his comments by saying that given the "National Security" implications of hafnium (the capital letters were his), he'd have to ask DARPA and the Department of Energy for permission to speak.

A few weeks later, he wrote to me again to tell me that I would soon get news from "DARPA folks" about how the final experiment was a "major success and that it matched the theory predictions I presented not too long ago.

"Other than that I do not have permission to continue the dialogue," he added.

But William Happer, a Princeton physicist and member of the JASONs' panel that had reviewed hafnium triggering, told me in the fall of 2005 that he had recently run into Khan at a cocktail party in Washington. Khan, who was trying to lobby the JASONs to support another hafnium experiment, also told Happer that they now had "irrefutable proof" of triggering, and were preparing to revive their funding request. "The story is not so good," Happer said. "It just shows how hard it is to stop a program like that when you have vigorous believers in Washington.

"You can keep it levitated for years," he joked.

James Carroll, Collins's former student, was still at Youngstown University and still being funded by the Air Force and the Army to do "isomer triggering" experiments. But rather than hafnium, he was working with isomers of americium and lutetium. Some scientists still bristled over Carroll's use of the word "triggering," and suspected him of keeping the intrigue over isomers bombs alive to perpetuate his own funding, but none at least were criticizing his physics. Carroll was glad, he told me, to be out of the hafnium business. Things were much calmer now and he was busy working with his students. He sounded genuinely happy.

I never met Fred Ambrose, the CIA/DIA official who had obsessively followed isomers for over a decade. When I e-mailed Ambrose, I got a testy phone call from a CIA spokeswoman, who instructed me to submit my questions to the agency's public affairs office. I did, and a few days later she replied that the CIA did not wish to comment on isomer weapons.

Paul Robinson, cochair of the hafnium production panel and the director of Sandia lab, left his position in 2004 to join Lockheed Martin Corporation in its bid to takeover the Los Alamos National Laboratory. With Lockheed tipped to win, scientists at the laboratory, currently run by the University of California, agonized over what life would be like working for a defense company, and colleagues at their sister lab, Lawrence Livermore, harbored similar concerns. Would a laboratory run by a defense company allow the sort of academic freedom and scientific integrity that propelled those who opposed the hafnium bomb. Not likely (though Lockheed ultimately lost).

And of course, Donald Rumsfeld, who was said to have wanted a hafnium bomb prototype in eighteen months, was still comfortably situated in the Pentagon's E-Ring, running the rapidly deteriorating occupation in Iraq. Eighteen months to build an exotic new weapon based on unknown physics indeed proved too ambitious for a military that was still struggling to provide basic armor to protect its troops from homemade bombs.

Another possibility suggested to me was that hafnium had "gone black," or in other words, had been slipped into the billions of dollar kept under wraps in the Pentagon's classified budget. One of the Los Alamos scientists had long suggested that in the 1990s—even prior to the 1998 dental X-ray experiment—Collins's work had been supported as part of such a "black program." Like the CIA psychics, such scientifically dubious programs, rarely scrutinized, can go on for years.

I guess we'll never know.

And that left me with one other mysterious question. I never did figure out how Payne Harrison, the Tom Clancy of isomers, had so accurately predicted DARPA's "isomer bomb" over a decade before the agency ever got involved with Collins. Dumb luck? I doubt that. I suppose there's no way to know, but I always suspected that somehow DARPA's connection to the isomer bomb long preceded the 2003 meeting when Stickley showed his infamous depiction of the hafnium hand grenade.

Or maybe not.

How does the isomer bomb end in Harrison's book? It's not a happy ending, certainly. After approaching the brink of world war over precious isomer reserves, the United States detonates a nuclear weapon—dubbed the Ostrich Egg—on the only known reserves of rubidium, vaporizing a bunch of poor Soviets who had just laid claim to the isomer. A rather extreme ending, although it did lead to a Soviet-U.S. treaty promising never to weaponize isomers.

If only real life were so neat.

What about Tony Tether's warning about massive Eastern European hafnium bombs raining down upon the United States, or slipped into suitcases bound for helpless American cities? Well, at least in Romania, Ioan-Iovitz Popescu, Collins's original collaborator, seemed to have moved on to what he described on his personal Web site as a "new explanation of the Newtonian law of gravitation." Popescu now believed the universe was filled with "etherons," a sort of modern spin on the long-rejected nineteenth-century notion of ether. He had touchingly added to his Web site a picture of his beloved wife's grave, which was engraved in English with: "4X Nominated for the Nobel Prize in Physics for Her Discovery of Multiphoton Spectra."

Of course, the controversy surrounding the isomer bomb did have sort of an accidental benefit for another farfetched weapon. Antimatter, or "positron," proponents could now claim that an antimatter bomb was far superior to the hafnium bomb. "Since nuclear isomers comprise long-lived radioactive materials, they also present the threat of collateral damage to humans and the environment," one private company wrote in their funding proposal to the Pentagon.

"Only positrons are deployable in both remote and confined areas, can be turned on and off at will and can be used without threat of contamination of humans or the environment (no collateral casualties)."

So, even if hafnium was having problems, the antimatter bomb was alive and well.

As for those on the opposing side of the hafnium gap, well, it's hard to say if their story ended happily. The JASONs, the group of eminent scientists that criticized the hafnium bomb, are still advising the Pentagon, but according to those familiar with their activities, the rift with DARPA runs deep. Peter Zimmerman, the former arms control official who first introduced me to the hafnium bomb, now teaches in London. William Herrmannsfeldt was, when last I spoke with him, enjoying his semi-retirement. He joked that no one was likely ever to appoint him to another panel, but he didn't care.

Some scientists complained that one of the real tragedies of the hafnium caper was that it made all isomer research "radioactive," so to speak. Meaning no one wanted to touch an isomer research proposal, given the hysteria over isomer bombs and the concerns about promoting a new "cold fusion." Any funding for isomer research—even legitimate investigations—was hard to come by after those halcyon days of hafnium. Perhaps in the end, it was science that suffered the most from the DARPA program.

Mort Weiss, the physicist from Livermore, seemed resigned to the fact that the hafnium bomb would probably continue in some form or another, just as he had predicted to me back at the beginning of 2004, when I first spoke to him. I once asked Weiss what his old friend Edward Teller would have thought of the hafnium bomb. Over the years, Teller in popular culture has been so closely associated with the hydrogen bomb and the overused image of *Dr. Strangelove* that it's often forgotten that Teller was, among other things, a brilliant physicist. Teller, on the other hand, was also called a "nuclear optimist," and entertained more than a few junk ideas. Even of cold fusion, Teller was known for making positive comments. Wouldn't the formidable Teller, if he were alive today, at least support the idea of looking into an isomer bomb? Weiss disagreed.

"When the scientific basis of isomers was less well understood, at least by me, Edward was enormously intrigued and supportive and much enjoyed the nuclear physics we were discovering," Weiss told me. "He understood why the physics we eventually had uncovered did not support more work in this area although he made amusing suggestions and accepted when I refuted them."

So, what would Teller have thought about the Pentagon's hafnium bomb, I asked?

Weiss replied: "Edward loved wild ideas but he was implacable in his understanding that physics was a science, not wishful thinking."

THERE WAS JUST one matter to resolve: the "final" TRIP experiment, the definitive proof that hafnium triggering existed? If successful, DARPA and the hafnium supporters could use the results as a way to revive the hafnium bomb. It could fail, but then, it seemed unlikely that devotees of hafnium would ever admit it failed, and how could it, if indeed the whole controversy hinged on the fact that there had never been a "success"?

Loyal hafnium-follower Patrick McDaniel had gone ahead and conducted his experiment. After being turned down at Berkeley, McDaniel conducted his final hafnium triggering experiment in early 2005 at Brookhaven National Laboratory. (As this book went to print, McDaniel had still not published any of those results, nor had he, to anyone's knowledge, even submitted those results for publication, although Ehsan Khan, in an e-mail to me, claimed the experiment was a success.)

At a fall 2005 meeting in the Washington area with the

hafnium clan to discuss the results of the final experiment, McDaniel apparently hedged. Maybe he saw something, maybe he didn't. Maybe there was a signal, or maybe it was just noise. Just like life. He needed, he said, more time, more money, more experiments. (Later, I was told that the hedging morphed into claims of "3 percent triggering," and shortly after that, DARPA's spokeswoman told me that McDaniel had seen 5 percent triggering, with a high degree of confidence—allegedly clear proof of triggering.)

Also present at that Beltway meeting, was Carl Collins, father of the imaginary hafnium bomb. Collins, too, had conducted a final experiment at the Japanese synchrotron facility, SPring-8. Collins's results, if successful, would surely provide some further fodder for the hafnium believers. But those results apparently weren't discussed. Did Collins at some point realize his results were wrong? Was Collins a true believer, a manipulator of lesser men, or just a sad dreamer and mediocre scientist, whose fading grasp at greatness led him to chase an imaginary weapon deep into the Pentagon's scientific underworld? Or maybe, as one of his opponents had suggested, his day of reckoning had finally come. In the final analysis, there's really no way to know.

But sitting in the meeting that fall day of 2005, Collins, for once, was uncharacteristically quiet.

SHARON WEINBERGER,
Washington, D.C.
December 31, 2005

AUTHOR'S NOTE

I ASSUME AN author's spouse is the most often sited acknowl-edgment for any book, and this will be no exception. My husband Nathan lived and breathed hafnium, the main topic of this book, for over two years, perhaps almost as much as I did. Living with an obsession is never easy, but living with the obsessed is that much harder. Nathan not only tolerated me during this time, but also made me feel normal.

Equally important, he edited every word of this book, helped to clarify my writing, and guided me back down a productive path when my mind wondered to other pursuits. This book, let alone the original article it was based on, would never have seen the light of day were it not for Nathan.

Thanks also go to my agent Michelle Tessler, at the Tessler Agency, whose rigor, patience, and belief in my work made this possible. Similar thanks go to Carl Bromley, editor at Nation Books, for his support and enthusiasm for this project.

I thank my friend Robert Wall, of *Aviation Week & Space Technology*, for his honesty both in criticism and praise, and for letting me know when I've colored a bit outside the lines. I'm also grateful to John Robinson, the editor of *Defense Daily*, who allowed me the time to pursue my pet interests. Thanks likewise go to Amy Butler, also of *Aviation Week*, and to Ann Finkbeiner, my professional confidante for all things related to science and society.

There is no way to thank all of the scientists who spent

countless hours answering my questions, by phone, over e-mail, and in person. I hope writing about them in the pages of this book shows my gratitude. Necessary to single out are Mort Weiss, Peter Zimmerman, William Herrmannsfeldt, and Jerry Wilhelmy. While this is not a book about the science of isomers as much as about the politics of science, I still hope that scientific readers do not find too much fault. Any mistakes that remain are undoubtedly my sole responsibility.

I thank Tom Shroder, editor-in-chief of the *Washington Post Magazine*, where the hafnium bomb was first published under the name, "Scary Things Come in Small Packages." Tom took the story I submitted in late 2003, called "Hafnium Dreams," and helped guide it into a coherent article. Many thanks also go to Susannah Gardiner, former managing editor of the *Washington Post Magazine*, who edited two more of my "fringe science" articles. Her insightful comments and questions helped improve and clarify my writing on these subjects.

Perhaps unusual for a book about fringe science, I also need to thank some of the scientists exiled—by choice or necessity—to the underworld. Many agreed to speak to me, fully understanding my skepticism of their work. Though I rarely agree with their conclusions, I am somehow drawn to their vision of a world that should, and perhaps even possibly could be, something other than what it is today.

In that vein, I thank Michael McKubre, a cold fusion scientist who has persisted for almost two decades in attempting to bring his work out of the underworld and into the mainstream. Cold fusion takes a bit of a beating in this book because I firmly believe that science is a process, not prophecy. Whatever the case, I do believe Mike's heart is in the right place and I wish him only the best.

Finally, some acknowledgement, though perhaps not exactly thanks, must go to Carl Collins, the anti-hero of this book. For over eight months we corresponded almost daily as I attempted to understand why he chased a dream that eluded him for so many years; a delusion that eventually drove him over into the very depths of the scientific underworld. Though we stopped speaking in mid-2004, I still can't help but feel that I learned some tiny, possibly sad lesson from Carl. Vision, even when etched with flaws, is poignant, and perhaps even more so for all its blemishes.

It's that fuzzy line we all tread between heartfelt conviction and obsessive delusion that made this long trip into the world of fringe science so intriguing. Obsession is tedium, but the decline into that obsession is an intricate mosaic of conflicting emotions and rationales. This book is about a journey through a scientific underworld, but it's also a portrait of human failing—something to which none of us are immune.

INDEX